Running Your Own
CATERING
COMPANY

Judy Ridgway

A catering business usually starts with a good cook – or someone who appreciates good food – who wants to do the job professionally. Such a business can be run on a part-time basis, at least to start with, and this has its attractions for those who want to launch themselves gradually.

This guide offers practical advice and information on how to cater for all types of function, from wedding and anniversary parties to business catering, drinks parties and children's parties.

The stringent hygiene requirements for food handling are set out, as well as the need to sort out legal requirements, location and premises, raising finance, buying the right equipment and employing staff, market research, advertising and public relations. The text covers:

- Choosing a catering activity
- Menu planning and costing
- Catering for different occasions
- Dealing with clients
- Setting up the business systems
- Adding a restaurant

The Author
Judy Ridgway is an experienced freelance author and feature writer on food, wine, cookery and catering, who has broadcast on radio and television as well as preparing a video on healthy eating.

Running Your Own
CATERING
COMPANY

Judy Ridgway

SECOND EDITION

KOGAN
PAGE

First published in 1984 entitled *Running Your Own Catering Business*
by Ursula Garner and Judy Ridgway.
Second edition 1992 by Judy Ridgway.

Kogan Page Limited
120 Pentonville Road
London N1 9JN

British Library Cataloguing in Publication Data

A CIP record for this book is available from the British Library.

ISBN 0-7494-0830-8

Typeset by DP Photosetting, Aylesbury, Bucks
Printed and bound in Great Britain by
Clays Ltd, St Ives plc

Contents

Introduction

Small catering businesses have been sprouting like mush-rooms in recent years and, despite the recession, many have prospered and some have grown into quite large outfits. Many who have come into the catering industry have been untrained either as professional chefs or as business men and women, but they have been good cooks, often with some first class specialities, and they have been keen to offer quality food backed up by friendly and efficient service. The formula certainly works at the small scale end of the catering market.

This type of firm generally works as a one-man band or as a partnership, backed up by teams of freelance kitchen and waiting staff. These are the small businesses referred to throughout the following text. Slightly larger or medium-sized catering companies may employ anything from two to a dozen or more full-time staff. Most of these will usually be in the kitchen or administration, with freelance back-up at the event. There are, of course, also some very large organisations who employ many more people and operate on a national or semi-national basis.

The progression from a small to a medium-sized company may come about as a natural result of increased business. However, it is far better to plan such expansion, particularly as it may mean a move to new premises and thus greater financial burdens, as well as an increase in the number of full-time staff. The latter also brings problems of supervision and maintenance of standards. The steps to even greater expansion need to be even more carefully thought out. Will the extra business come from increased volume in the same areas or will you need to diversify into new areas? Is the timing right both financially and in regard to the market place? Can your existing management cope with the extra volume of business? These are all questions which will need to be answered.

The scope of work carried out by catering companies is

certainly wide enough to allow for more entries into the field but you will need to study your own area in some depth to see what kind of services might do well and whether or not they are already being offered by competitors.

Potential activities in the private sector take in every kind of function from christening parties to funeral teas and everything that happens in between! The list must include children's parties, birthday parties, engagement parties, weddings, anniversary parties, general drinks and buffet parties, retirement celebrations, luncheons, dinners, picnics, barbecues, brunch parties and late suppers and you can probably think of another dozen.

On the business front there are opportunities for catering for every type of meal from the desk-top lunch to a full banquet. There will also be buffets, brunches and maybe even breakfasts. Then there are group activities, such as club dinners and dances, large-scale barbecues and group travel hampers.

Some companies have built up their business by concentrating on one or two specialist services such as desk-top dining and picnics while others offer a comprehensive service. Some also combine their catering operations with related services such as dining suites, frozen foods supplies or delicatessen shops.

Of course, thinking about running a catering company and actually doing so are two very different things and you will need to be very sure that you have what it takes. The first question is, how do you feel about hard work? You may reply that you are quite used to that but the truth of the matter is that hard work for an employer is nothing compared with hard work for yourself. You could be getting up at six in the morning to batch bake for the next day's wedding, spend the morning preparing a small buffet lunch and then supervising the service. This may be followed by a small cocktail party in the evening which drags on till nine or ten and then there are the records and accounts to fill in on those and yesterday's jobs.

In the early stages the chances are that you will also spend considerable time humping large bags and boxes of food, drink and equipment, from the cash and carry to your kitchen, from your kitchen to venues which will inevitably be on the second or third floor and thence back home again. So you must be physically strong and you will certainly need to have good health.

If you are not daunted by hard work then you should cast a look at your social life, for this could be the last you will see of it for some time. If you are involved in private catering you will find that, naturally, everyone wants to entertain in the evenings and at weekends. This will intensify on bank holidays and at Christmas and Easter. Even if you concentrate on dealing with business catering, you will still have the books to do in the evenings and batch cooking at the weekend, and the chances are that you will also be taking on the odd weekend wedding or large evening party simply because it can be lucrative.

The next question must be, how does your partner or family view the proposed venture? They will be the ones who will bear the brunt of the change in your social life and they could also be affected by the increase in kitchen use, your lack of time, and the need to steer clear of your office area.

Once you have faced these initial stumbling blocks, take a cool look at yourself and your personality. How are you at dealing with people? Your staff are essential to you and you must be able to get on with them, direct and support them. Your clients, too, must find you equally friendly, helpful and efficient.

Are you completely unflappable – even if the glasses have not turned up, the ice is melting before you can properly cool the champagne, and one of your staff has just dropped a whole tray of canapés? How resourceful are you? Will you know how to roll out a batch of pastry in the absence of a rolling pin or open a beer bottle without an opener?

Having decided that you can cope with most eventualities, how much do you know about the catering business? If the answer is 'not very much' it might be worth working on a freelance basis for another outfit. This would certainly give you some experience in cooking and presenting the food, and if you keep your eyes open you should be able to learn quite a lot more.

There are also a number of courses which you might consider if you have the time and money to take a break between your current work and starting up your own business. These courses may cover the cooking (see Appendix) or they may cover the business side of your proposed activities. You will know which will be the most valuable to you.

When you have completed this kind of personal appraisal and perhaps undertaken some preliminary training, you will be in a much better position to take the decision on whether or not to go ahead with launching a new catering service. And once that decision is taken the groundwork can begin.

Chapters 1 to 11 look at the decisions that must be made before you can really get off the ground. Practical advice on the organisation of different kinds of functions follows in Chapters 12 to 18. Chapters 19 and 20 look at how to get your services known and how to build up a good reputation. Chapter 21 looks at the possibility of adding a restaurant to your business.

Chapter 1

Business Opportunities and Market Research

It is, of course, quite possible to set up as a caterer without any clear idea of what kind of business you are looking for and many smaller businesses have started off just like this. Such caterers have been prepared to turn their hands to anything, and slowly the business has trickled in and experience has gradually pointed the way to the more profitable business opportunities. But this is at best a rather haphazard method of going about things and at worst a recipe for low income, cash flow problems and possible bankruptcy.

Far better to assess all the possibilities in advance and to check out the most promising against some careful market research and your own inclinations. The latter consideration can be more important than sometimes given credit for. There is, after all, no point in setting up a dinner party service when you would really prefer to avoid evening work. Things could work out to start with, but you may find that as time goes on you lack the enthusiasm to keep the initial impetus going.

On the other hand, a dinner party service might fit in with your particular life-style very well indeed. But if your catchment area is largely inhabited by 30-year old couples with young children and fully stretched budgets you will have to think again. Equally, a popular retirement area will not be the place to start up a children's party service, and a small market town is unlikely to support a business based on desktop lunches.

A detailed look at the potential market will help you to put together a much more viable business plan than would otherwise be the case. Such a plan, supported by detailed analysis, will be essential if you are thinking of trying to raise outside finance. It will also be an extremely useful guide when taking decisions on subjects as diverse as the type and amount of extra kitchen equipment to buy, putting together menus and prices, and planning a publicity campaign.

Catering activities

The best way to start a business plan is to list all the activities which a catering business might reasonably take on. Work with the checklist below. Expand sections as ideas occur to you and ignore any market information you have at this stage. It could inhibit your imagination, and the exercise is really a kind of brainstorming session which might throw up some unusual and possibly profitable ideas for the future. You can sort out the practical suggestions from the flights of fancy later on. All the activities mentioned are covered in greater detail in subsequent chapters.

Domestic market
The activities listed under this heading will all be intended for private individuals or groups of people organising their leisure pursuits.

1. *Dinner parties*
A dinner party service can work extremely well in a highly residential area with a reasonable level of wealth. Requirements will vary from those hostesses who require a complete service, including waiting on and washing up as well as shopping and preparation, to those who want only to have the food delivered so that they can pass it off as their own.

Numbers may vary from four to 40 but you will need to be aware of the pitfalls at both ends of that scale. Unless you have a specially adjusted price structure, four people can be quite uneconomic. On the other hand, 40 may present problems in the size and capacity of your cooking equipment and extra staff will probably be required.

Depending on the spending power of your potential customers, the business may be scattered among a large group of customers who will be using your services only for special occasions, or it may be confined to a much smaller group of regular clients. The latter business can spill over to orders to stock up for country weekends. There may also be some call for catered luncheons from such clients.

Jean Pierre Despesme of Jean Pierre's Pantry started with just such a group of regular and wealthy clients, and his frozen food business grew up in response to requests from customers for food similar to his dinner party offerings to take to their weekend residences. He gradually stocked up his first freezer by cooking between catering engagements to fill in time.

2. Buffet parties

This is another reasonably profitable area. A good many people will occasionally bring in outside caterers for their parties and if your catchment area is a good one, you could specialise in buffets. As with dinner parties, hostesses vary in their requirements and the demand may be for a complete service or simply for the delivery of the food. Numbers may vary from 12 or 20 to as many as 100 or more and you will certainly be involved with extra staff and catering equipment hire. The buffet may be finger or fork food.

3. Cocktail parties

These are unlikely to form the full basis for your activities but they can be extremely profitable. However, bear in mind that canapés can be extremely time-consuming to prepare and must be priced accordingly. Numbers may vary from 20 to 200, and if you are providing a full service, extra staff will be required and special care will need to be taken with the drinks.

4. Weddings

Wedding breakfasts may take the form of a buffet or canapé party or they may be sit-down meals but, because of the special considerations involved, they do form a separate area of activity in their own right. As caterer, you need to be an expert in wedding protocol, for you will often be called upon to advise the bride and her family. In the right area weddings can be a lucrative speciality. Numbers may vary from 20-odd to 200 or more. In addition to the usual requirements, you may need to deal with marquee and furniture hire, provide the cake and act as master of ceremonies.

5. Anniversaries and special occasions

Like wedding breakfasts, these may take the form of a dinner, a buffet or a cocktail party, but by highlighting your expertise they can become an important profit centre. Janet Raglan's Catering Company is based in a wealthy retirement area and she has made a speciality of wedding anniversaries from pearl (30 years), through coral, ruby and sapphire to gold (50 years) and emerald and diamond.

The same idea might be exploited for younger age groups in other areas. Birthday parties, particularly for the decade, and retirement parties are other ideas. There may also be a call for catering specialised celebrations such as bar mitzvahs, coming of age parties or Hogmanay.

6. Children's parties
Children's tastes are quite different from those of adults and a successful party service could be inundated with business. Many mothers will say to a first time party-giver quite simply, 'Don't!' But if the preparation of the food and the organisation of the party are undertaken by a caterer, the mother has a much easier time. Numbers could vary from 12 to 20 – resist too large a number or it will be chaos. You may need to be able to organise games and entertainments as well as provide food, cake and presents. Theme parties are popular with children so make sure there are plenty of creative ideas in your brochure.

7. Outdoor parties and picnics
British weather being what it is, these are only likely to supplement your business during the summer months, but barbecues are growing in popularity and a specialised service in this area could catch on. All outdoor parties pose special problems but they can also offer something a little different.

Picnics and hampers can be a useful adjunct to your business, particularly if you are situated near famous beauty spots or the venues of well-known sporting events, such as race courses. One-off offers for annual events are also worth considering.

8. Specialised food catering
There are a small number of businesses which have made a success of specialising in wholefood or vegetarian menus. However, unless you are sure of your market, it is probably more sensible to offer such specialities as an additional service.

9. Catering for clubs and associations
Most residential areas abound with clubs and associations covering almost every type of leisure activity. Most of them have annual social events and many organise much more frequent get-togethers. A caterer specialising in just such

events could be extremely welcome. The chances are that there could also be a useful spin-off into private functions.

10. Frozen food service
This is a fairly specialised activity which is sometimes run on its own and sometimes in conjunction with a general catering service. Opportunities include freezer stocking and the provision of complete frozen dinners and buffets. However, the larger operators offer a range of single portion dishes which can be purchased individually or in bulk.

The outlay for this type of operation is likely to be much higher than for a small catering business which could be started with minimum capital. Frozen food requires a considerable investment in freezers and bulk preparation and cooking equipment. New or extended premises will probably also be required both for the preparation and packing of the food and for the sales operation. Additional sales outlets will need to be ferreted out and a continuous supply of orders maintained.

11. Delicatessen caterers
This is another logical extension of the idea of catering from special premises and quite a number of such shops have opened recently. The food is all prepared on the premises and sold both in the shop and through the catering business. The shop acts as a sampling operation as well as a retail outlet in its own right. Customers can choose the degree of service they require, coming to the shop or having the food delivered to their homes. Service and special menus will also be available.

12. Other catering activities
Once you have your own premises, the possibilities start to open out but, of course, so do the problems. Hiring out premises for social functions, for example, can work well in an area which has little in the way of function rooms other than the local hotels. Demonstration kitchen facilities could find a ready market in the major cities where PR activity is high. Training and cookery courses offer other possibilities.

Commercial market
These activities will all aim to provide a service for businesses in your catchment area. Some will be similar to those offered to the private market but there are differences in timing and

organisation. Most business activity is likely to be at lunch time and on weekdays, whereas private functions tend to be in the evening and often at the weekend. Methods of payment will be another difference. Private clients will usually pay on the day or shortly after receiving your bill. Business customers are much more likely to sit on your invoice for a couple of months!

Business luncheons
This is likely to be by far the largest of all the commercial opportunities. Most business entertaining is done at lunch time and this holds good for small gatherings and large press receptions or VIP visits. Business meetings often carry on through the lunch period and the participants still need to eat.

(a) *Sit-down luncheons*
These are usually fairly small boardroom affairs and you may need to work out a special price structure to ensure that they are profitable. However, larger numbers may be catered for at press functions and at Christmas time.

(b) *Buffet luncheons*
This is where the bulk of the business lies: conferences, press receptions, general receptions, annual, monthly and weekly meetings of trade federations and institutions, demonstrations, sales meetings and management/union meetings, openings and overseas visitors all provide the need for a meal. Buffets may be simple or extremely elaborate. Numbers are likely to vary from 20 to 200.

(c) *Cocktail and drinks parties*
These are more likely to be held in the evening but cocktails will be over by eight or nine o'clock. Small events may be held to mark presentations and retirements but larger events could run into hundreds. Evening drinks parties, possibly with a buffet, become much more frequent over the Christmas/New Year period.

(d) *Lunch and picnic boxes*
Lunch boxes or sandwiches and finger food may be ordered on a regular basis for office meetings and working lunches. They may also be ordered by senior

executives who do not want to go out for a meal at lunch time but who have no in-house catering facilities. Picnic boxes may be ordered for public relations trips to sporting events and the like.

Your skills and inclinations

The second area to look at in some detail is your own skills and inclinations. Here are some questions to ask yourself. The answers will help you to cut down the list of possible activities discussed above.

1. Do you have any special training?
Answers may include a full-time catering course, cordon bleu training, some evening classes or nothing at all. If you are a good cook it will not matter if you have no training, but you must be sure that you will be able to cope with large scale cooking and will be able to produce a particular dish which has been specially requested by your client. You must also be able to direct other cooks when catering for large functions or when your business expands. You may feel that you should add to your professional training and courses are listed in the Appendix.

2. Have you any special skills?
These may have been developed as a result of training courses or classes or they may just have been acquired over the years. Obviously, a skill in pastry- or bread-making, cake decorating or decoration work could point you towards a preference for particular activities. However, special skills are not essential. It is usually more useful to be a good all round cook. You can always buy in special expertise.

3. Have you any special inclinations?
What kind of things do you enjoy cooking? However good you are at a particular culinary activity, you will soon become bored if you do not enjoy doing it. A dinner party service, for example, will entail a quite different type of catering from that required for desk-top lunches and picnic boxes.

4. What are your time constraints?
Are you prepared to work in the evenings and at weekends or

would you prefer to restrict business, as far as possible, to lunch times and weekdays?

5. How large an operation are you planning initially?
The answer to this may depend on available capital or it may be tied up with your decision to take in partners or to employ people.

Market research

The next stage in putting your business plan together is to study your catchment area. The more detailed this study is the more useful it will be.

Start by defining your catchment area. This really means deciding how far you are prepared to travel to fill your order books. In a large city you may be able to draw a line round a five- or ten-mile radius. However, your study should also take in the surrounding area, for you may find that the addition of a further five miles in one direction will bring you into a much wealthier area or will enable you to consider business luncheons.

Most caterers running businesses based outside the large conurbations accept that, from time to time, they will probably have to travel quite considerable distances.

Next, try to assess your catchment area in terms of life-style, tastes and spending power. Here are some of the factors on which you should check:

Local population trends
Is the population expanding or decreasing? What is the average age of that population? Figures for the past five years should show the trends. Check your local council for figures or ask your local library to help you to track them down. This information will help you decide if the area is predominantly a retirement area, a haven for the wealthy middle-aged or simply full of young couples with children.

Income trends and wealth
What is the average income in the area and thus what kind of spending power? Assess this by driving round the area to look at the property: talk to estate agents and see what kind of shops there are in this area and the value of the merchandise they offer.

Group activities, clubs and leisure complexes

A look at the kind of local activities that are available will help to reinforce your conclusions on life-style and spending power. It may also point to specific markets. Check if there are any sporting venues such as race courses or sports stadia. Is there a permanent show or exhibition site?

Business and industry

What is happening to business in the area? Is it expanding or contracting? This could be important both for private and business functions. New industrial estates will mean a larger number of executives moving into the area. They may also mean a greater potential for business entertaining. Are there any really large firms, government departments or large insurance companies in the area? All these may generate business. Talk to local planning officers, check the Yellow Pages and the phone book.

Competition

It is also important to look at the potential competition. Their activities may be useful in assessing the market. There may be gaps in their repertoires which you could usefully fill. Check the Yellow Pages and any other local directories and walk round the area to look at delicatessen and specialist shops.

Checklist of potential business

From the above research you should be able to make a list of different categories of potential customers, together with an indication of the kind of services each group might be interested in. Here are a couple of hypothetical examples.

The first covers a large seaside town with a large retirement population but also with a wealthy residential area to one end and a thriving tourist trade. Leisure activities include a busy sailing club, a well-known operatic venue and a large sports complex. There is also a small industrial estate and the head office of a major distribution company.

Retired population. One-off anniversaries and family gatherings.

Wealthy inhabitants. Dinner party service and larger parties. Hampers and picnics.

Sailing, tennis, cricket and bowling clubs. Club functions.

Operatic venues. Hamper service for the season.

Tourist track. Special hamper service through hotels which do not have a large enough staff to fulfil demand.

Industrial estates and headquarters of distribution company. Business lunches, desk top and buffet, for meetings and for entertaining. One-off functions at Christmas.

The second example covers a residential area about seven miles from the centre of one of Britain's largest cities. The local population is fairly young but with a reasonable income.

Business community. All kinds of business functions including desk-top lunches, buffets and cocktail parties and PR events.

Local population. Children's parties. Buffet parties and barbecues. Club functions.

The opportunities here are very much greater than in the first example and the decision to concentrate on business or private functions will probably depend more on your own inclinations than on the potential volume of work. Once that decision is made you should go back and look at the relevant markets in more detail. In the seaside town, on the other hand, it may be necessary to explore all avenues of potential business before you are in a position to decide which to concentrate upon.

By working with the above checklists you should now be in a position to write out a fairly realistic assessment of the pattern of potential business for your firm and from this a statement of exactly what you are planning to do.

Chapter 2
Finance

It is possible to start up a small catering company with a relatively small amount of capital. Your own kitchen may possibly be suitable for use and you can hire any special equipment. Extra staff can be used on a spasmodic and freelance basis and changes can be made as you gradually plough profits back into the business. All you would need is a small amount of working capital to finance your first two or three months' trading.

However, if you are really serious about the business you will probably need to make a greater investment than this. The sums involved will depend upon the size of business operation you have in mind and upon the type of catering activities on which you expect to concentrate. The business statement outlined in the previous chapter should provide the guidelines here.

Estimating the finance required

To estimate the amount of money you will need, check through the following areas of expenditure:

Premises
Will your existing kitchen be suitable or are you planning to set up the business in quite separate premises? If a shop forms part of the business plan or if you are thinking of setting up demonstration kitchens or catering facilities for hire, then this will form the major part of the fixed capital requirement.

If you are planning to use your existing kitchen in your private residence, it may need to be extended or redesigned and refurbished to satisfy the requirements of the Food Hygiene Regulations (see pages 34-5) and this too could cost quite a lot of money.

Capital equipment

After the structural considerations, there is equipment to consider. New premises will, of course, demand new or perhaps second-hand equipment but it all needs to be purchased before you can start up.

Your own kitchen may need the addition of large cookers and greater storage space. You may have to buy new mixing and processing equipment, larger pans and special utensils or bakeware. Serving dishes will need to be considered and you may decide to stock up with sufficient plates, glasses and cutlery to be able to provide for small functions without hiring. See Chapter 6 for more details of possible requirements.

Capital equipment may also take in office equipment such as computer, answerphone, word processor, fax, and duplicating equipment.

Transport

It is almost impossible to run a catering business without some form of transport. This is the time to decide whether or not you need a new car, estate car or van, or whether you can manage with the one you have.

Starting-up costs

There are costs which will be incurred simply by starting up. These could include stationery and office materials, menus and price-lists, packaging materials and bulk purchase of groceries. Ask yourself what you will need for your own particular operation. If, for example, you are planning to specialise in hampers and picnics, you will require a supply of the actual hampers and picnic boxes to hand. A desk-top dining service will call for a good stock of paper plates, plastic crockery and paper napkins, and a speciality line in cakes and desserts will require cake boxes.

Starting-up costs will also include new telephone installation charges, insurance on the business and its transport, legal fees and publicity. In fact, everything which needs to be paid for before you start trading.

Working capital

Last but by no means least you will need the money to buy the food and pay the other bills incurred during your first few months of trading. To determine how much this is likely to be, you will need to work out an estimated budget and cash flow forecasts. These will support any business proposal you put together to back up requests for capital to banks and other financial institutions. In addition, they will be very useful tools in assessing whether or not your pricing system is right, and in seeing how your business is faring (see page 30).

Organising the business

As soon as you have estimated the costs involved in setting up, you will need to start talking to your bank manager, solicitor and accountant. But before you do this, you must also have some idea of what sort of business you are planning. What is its legal entity to be? You may be planning to go it alone or you may feel that the business plan you have in mind needs more than one person to manage and control it. You may even have prospective partners or directors in mind.

If you operate as a sole trader you can use your own name or a made-up name such as 'Best Bib and Tucker'. You do not have to register this name but you had better be sure that there is no other name close enough to accuse you of trading on their name. There are also some words such as Royal, English and Association which may not be used. You can get a list of these from Companies House.

If you use a name other than your own you must show your name and address as proprietor in your business premises and on your stationery. You do not have to submit accounts to Companies House but you will be liable to pay National Insurance and Income Tax for yourself and anyone you employ. The main disadvantage of being a sole trader is that you are personally liable for all your business debts.

A partnership is a firm with two or more proprietors. The same conditions apply as for sole traders. However, when it comes to liability for debts the law says that all parties are jointly and severally responsible. This means that if your partners abscond you will be liable to pay their debts as well as your own. Because of this you should have a legally binding partnership agreement drawn up by a solicitor even if you are going into business with people who are well known to you. The agreement should cover such topics as who signs cheques (at least two partners) and what happens if a partner wants to leave.

When you set up a limited company you are creating a new

legal entity. A company can sue and be sued just like a person. A company must have at least two shareholders, one director and a company secretary who can be a second director. The shareholders' liability for debt is limited but in practice banks and other lenders usually ask for personal guarantees against loans. Annual accounts must be submitted to Companies House and directors are subject to company law. There are also a whole host of other provisions and you would be well advised to talk to your accountant and solicitor, or read the relevant section of *Working for Yourself: The Daily Telegraph Guide to Self-employment*.

Your accountant and solicitor and, above all, your bank manager, are going to be extremely important to you both in the initial organisation and setting up of your business and in the running of it. If you already have good working relationships with such people it is sensible to stick with them. If you haven't, the choices will be important ones. Start by talking to other business people in the area and see who they are using. If you are already in touch with one local adviser or you know your bank manager well, ask him to make some recommendations. Get together a short list of suitable firms in each field and take your time about the choice. Do not simply go to the nearest. Go and see each firm on your list. Make a list of items to check and questions to ask. Be clear in your own mind about the areas in which you think you will need help.

Raising the money

The more money you can raise yourself the easier it will be to get going, so start by working out just how much you can put together from your own resources.

Possible sources of personal finance

1. Have you any spare capital in building societies and the like?
2. Have you any shares or securities which could be sold or used as security against a personal loan?
3. Have you just received redundancy pay or a retirement annuity?
4. Have you any jewellery, paintings or antiques which might be sold?

5. Have you a second home, a caravan or a boat which might also be sold?
6. Do you have any insurance policies against which a loan might be negotiated?
7. Is there a chance of a second mortgage on your house or could you sell up and buy a smaller, cheaper one? The latter might be the easier course if you are married and need your partner's agreement to a second mortgage.
8. Your business partners, if you have decided upon a partnership, may also be able to raise money in the same ways.

Depending on your circumstances and on how ambitious your business plan is, these methods may produce sufficient money to go ahead. If they do not then they may still provide enough to enable you to get a loan for the rest.

Bank loans

The bank is the most likely source for the money you need and may be prepared to offer a lump sum, usually equal to that which you have been able to raise yourself, towards any initial investment required. It may also offer an overdraft facility to help with the demands upon working capital. This is the commonest form of bank help and it has the advantage that interest is paid only on the actual amount you are overdrawn. The disadvantage is that an overdraft can be withdrawn at any time, so it should be used only to cover short-term requirements.

The bank may ask for personal guarantees for the money they are prepared to lend you. Larger branches are more likely to lend larger sums than the smaller branches but if you can get a small branch to invest in you there are advantages in the closer relationship you are likely to establish there.

Some banks run their own start-up schemes and some foreign banks are reputed to be freer with their risk capital. So shop around but keep your solicitor fully informed of your intentions and be sure they check the small print in any agreement.

The banks also administer the government Loan Guarantee Scheme. If you are eligible for this scheme the government will guarantee to the bank on your behalf 70 per cent of the amount borrowed. But, of course, you are still liable for repayment of the loan to the government. There is also an interest premium on the amount guaranteed. Ask for details from your bank or get booklet PL876 from the Employment Department.

Private loans
Very often the difference between the total of your own capital and that required to start up in business can be found within the family or from friends or ex-business associates who are prepared to back your ideas. The best way to arrange this is for your backers to guarantee a loan or overdraft on the basis that, if the recipient is unable to pay up, then the guarantor is liable for that amount. The advantage of this kind of arrangement is that, in certain circumstances, the guarantor of a business loan can treat any losses incurred as capital losses which can be used to offset capital gains. Details of the Business Expansion Scheme are available from your local TEC or LEC; the Scheme is due to be wound up at the end of 1993. The limit is £40,000 and close relatives are excluded.

The outright loan of capital sums can be fraught with difficulty, however. First, it must be made clear to what extent, if any, the lender has any say in the running of the business and what the nature of this control is. Such control is to be avoided if possible and the existence of a loan should not normally entitle the lender to any participation in management matters nor to a share of the profits.

Misunderstandings can also occur concerning the terms of the loan, and the best plan is to get a solicitor to draw them up formally, covering the rate of interest, the duration of the loan and how it is to be repaid, the circumstances under which the loan may be withdrawn and any other conditions.

Sometimes private loans, and indeed those from other outside agencies such as finance houses and investment companies, are conditional on acquiring shares or the option to acquire them. This means that you will be giving away part of the company and you are unlikely ever to get it back. The pros and cons of such a procedure are set out in a booklet entitled 'Finance Without Debt: a Guide to Sources of Venture Capital under £250,000' published by the Employment Department. It is also worth investigating the Share Buy Back Scheme under which it may be possible to sell shares on a temporary basis. Ask your accountant.

Other sources of finance

If you are planning a major operation with new premises you may need to look further afield for the larger amount of capital involved. Possible sources include regional development boards, the Rural Development Commission, British Coal Enterprise Ltd and British Steel Industry Ltd. *Working for Yourself: The Daily Telegraph Guide to Self-employment* contains a very useful section on finance.

If, on the other hand, you have a source of capital and just need a small amount to help you keep going one of the 82 Training and Enterprise Councils (TECs) may be able to help you with business advice, counselling and training as well as £50 per week for your first 26 weeks of trading. You must have been unemployed for six weeks and your business must be a new one.

It is well worth considering hire purchase agreements if most of your requirement is for a car or van or for equipment such as cookers or freezers. This is a useful way of financing medium-term commitments. The payment is spread over a period of time and, until the period of the agreement expires, the item remains the property of the finance company and the borrower does not have to provide security as is the case with a loan. However, rates of interest can be high and it is worth getting more than one quote if you can.

Putting together a business proposal

This does not have to be as complicated as it sounds, but it can be invaluable in showing that you have studied your market prospects in detail and understand how the proposed business should be run. The bank manager or investment company will be looking for an assessment of the business potential of the services you plan to offer, a detailed breakdown of starting-up costs and budget, cash flow and profit forecasts. What your proposals should show is that you understand the business you plan to enter and have a clear idea of how things will go. A typical proposal will include the following headings:

Personal and company details

Details of the proposed structure of your firm together with the names and addresses of the partners and directors, if any. It should include an outline of your own history and that of anyone else involved, with details of relevant experience and qualifications.

The business plan
This will be taken from the business plan worked out in Chapter 1. It will outline the type of work on which you expect to concentrate, together with any specialised sidelines which could add to the overall profitability of the business. This information should be backed up by the market research information you have gleaned.

Financial requirements
Details of all the money needed for the purchase of premises or capital equipment as well as that required under the headings of starting-up costs and working capital. It should be itemised in some detail and a further section may be required to give full details of new premises, planning permission, surveyor reports etc. This should be followed by a breakdown of the money already available from personal or other sources and that which is still required.

Sample menus and price-lists
These will include a detailed breakdown of how the prices are computed and an estimate of the volume of business required to break even.

Budget for the first six months of operation
The budget will include a month-by-month or week-by-week estimate of income and expenditure, the latter being broken down into fixed and variable costs. If business is expected to be seasonal, the budget may need to be projected for a year or longer. Notes on how the estimated income has been arrived at should also be included.

Cash flow forecast
This will cover the same income and expenditure information contained in the budget, but instead of being averaged out over the entire period, the sums will be shown as actual amounts in the months in which they fall. Starting-up costs may be included in the first month (see page 24).

£12.95 Paperback   ISBN 0 7494 0408 6   208 pages   1992

*Available from good bookshops or direct from the publisher (add 10%
p&p):Kogan Page, 120 Pentonville Road, London N1 9JN.*

References
This section will give names and addresses of financial referees.

The preparation of this business plan will clarify your ideas, and enable you to discuss all aspects of the project both fluently and objectively with potential lenders.

It is worth writing up such a business proposal even if you do not really need any outside finance. It will certainly provide a very good base from which to start. When you have written it, have another look at the financial requirements, and ask yourself, 'Have I covered every eventuality?' It may be sensible to add in a little more to cover something which may cost more than you have budgeted or which you simply have not thought of.

Next ask the complementary question: 'Do I really need to spend all this money at this stage?' This question is even more important if you are going to borrow the money. It is not wise to over-commit yourself. Far better to start off in a smaller way and build up the business over time than to suddenly find you cannot make ends meet.

Chapter 3
The Legal Requirements

Once you set up in business there is a mass of legislation which needs to be taken into account. It covers subjects as diverse as the hygiene of your kitchen, the details printed on your packaging, the way you treat your employees and the records you will need to keep. Some of the legislation could affect you even before you start trading. You will, for example, need to have planning permission from your local authority, and the Environmental Health Officer will want to inspect your kitchen.

Most of the relevant Acts of Parliament are pretty complicated and you may need to take expert advice in various areas. However, some government and local government departments issue booklets explaining what the requirements are and how they are likely to affect particular businesses. Copies of the Acts themselves can be obtained from HMSO headquarters and HMSO bookshops and a study of these, together with any explanatory booklets which are available, should point you in the right direction.

The various areas affected by legislation are:

The premises

In theory, you need to have planning permission for any kind of change in the use of a building, and if you are planning to run your business from home this could be seen as involving a change of use for at least part of the time. In practice such a change may not be evident from the outside and will probably not result in any inconvenience to the neighbours or any change to the nature of the area. However, the local council could order you to stop trading from your house if they objected to your activities.

The need for planning permission becomes much more acute if you buy or rent property outside your home. If the property has previously been used for other purposes you must get planning permission from the local authority for a change of use. This holds good even if the premises have been used as a shop or restaurant before. However, the nearer the previous use to your current proposals, the more forthcoming permission is likely to be.

Whatever the previous use, it is only sensible to plan your application in some detail and research it well. The Planning Committee will want to know exactly what you are planning to do, how these plans will affect the immediate neighbourhood and what demand there may be for your particular services. Make sure that you include structural plans and detailed drawings with your application and do list absolutely everything that you hope to do to the building.

Extensions to your existing property must also be submitted for planning permission and you would do well to check with your local Fire Officer and Environmental Health Officer before submitting your plans. There may be some requirements you have not taken into account. The Occupiers Liability Act, for example, covers the duty of the occupier of premises to visitors to those premises, and the Health and Safety at Work Act makes provision for good lighting, ventilation, toilet facilities and the like for employees.

The Food Safety Act 1990 and the Food Hygiene (General) Regulations are probably the most important legislation in this area. The Act requires you to register with your local authority the moment you make food for sale. This means contacting your local Department of Environmental Health. The Food Hygiene regulations take in all aspects of the cleanliness of your premises and cover the structural make-up as well as the kitchen surfaces, ventilation, washing facilities, equipment, drainage, storage and refuse. Some councils issue useful guidelines or notes on their requirements and the inspectors are usually extremely helpful when it comes to planning new premises or making changes to existing ones.

Checklist of relevant legislation
Fire Precautions Act 1946
Food Safety Act 1990
Food Hygiene (General) Regulations 1970
Food Hygiene (Amendment) Regulations 1990 and 1991
Health and Safety at Work Act 1974
Occupiers Liability Act 1957

Offices, Shops and Railway Premises Act 1963
Shops Act 1950
Smoke Detection Act 1991

The food

Once you have got your premises sorted out and start to plan your menus, or the food to offer for sale from the freezer or over the counter, you will come up against the next batch of legislation.

Some foods such as sausages, meat pies etc have very strict standards for their composition, and if you are planning to sell them you will need to check with the regulations. However, most of these refer to the meat content levels which are so low that any quality product is likely to be way over the specifications. The use of preservatives in food is also regulated and this may also affect your operation.

If you are planning to include a frozen food service in your activities then the Food Labelling Regulations will come into force. This legislation requires quite a lot of information to be included on the label. You must, for example, state the name and address of the maker, a list of ingredients in descending order of weight, the precise name of the food, an indication of shelf life and how it should be stored, and any special instructions for use. All this information must be easy to read and easy to understand.

Claims made for your produce must be within the limits of the Trade Descriptions Acts, and you will also need to have a working knowledge of the Weights and Measures Act.

Checklist of relevant legislation
Food Labelling Regulations 1980, 1984
Food Labelling (Amendment) Regulations 1989 and 1990
The Materials and Articles in Contact with Food Regulations 1987
Meat Products and Spreadable Fish Products Regulations 1984, and Amendments 1986
Miscellaneous Additives in Food Regulations 1980, and Amendments 1982
Preservatives in Food Regulations 1979, and Amendments 1980, 1982
Trade Descriptions Acts 1968, 1972

Weights and Measures Regulation 1985 and Amendments 1986, 1987 and 1988

Running the business

The Sale of Goods Act and legislation dealing with the hygiene of delivery vans will affect you when the planning stages are over and you start to trade.

You will also become liable for income tax and, possibly, corporation tax. Value added tax will also have to be paid. Unlike income tax, it is calculated on the turnover of your business rather than on the money you make after paying all the costs. The VAT limit is currently set at £36,600 (1992 figure) per annum and, unless your business is very small indeed, you will be liable from the start. You register for VAT at the local Customs and Excise office and you should find the VAT inspectors very helpful. If you are in any doubt about how to carry out your obligations they will usually arrange to visit you.

If you are employing people, even if only on a part-time basis, you will also be liable for organising their income tax deduction under the PAYE scheme and for National Insurance contributions both for yourself and your staff. Employing staff will also involve you in the mass of employment legislation covering everything from conditions of dismissal and redundancy, through accidents and disease reporting, to race and sex discrimination. The Inland Revenue and the DSS both publish leaflets for employers.

The way round some of this legislation is to employ only casual labour on a one-off job basis. The individuals so employed will be responsible for dealing with their own tax and National Insurance but you are still liable for the standards of work under the hygiene regulations and for the provision of suitable conditions of work.

Checklist of relevant legislation
Materials and Articles in Contact with Food Regulations 1987
Employers Liability Act 1969
Employment Protection (Consolidation) Act 1978
Employment Acts 1980 and 1982
Equal Pay Act 1970
Fair Wages Resolution 1946

JOIN
THE HCIMA

How can you benefit?

Because we're here to help you.

Indeed, the HCIMA – with over 24,000 members, worldwide – is the only internationally recognised professional body representing and promoting the interests of managers in food and accommodation services.

Whilst playing a vital role in establishing and helping to maintain quality standards throughout the industry, the HCIMA offer members comprehensive support and a wide range of dedicated services, including:

- Modular professional qualification programmes.
- Comprehensive computer-based information service.
- Professional advice on career enhancement, job opportunities, legal and technical concerns.
- Relevant seminars on a wide range of industry issues.
- Special discounts and exclusive offers.

To find out more about the HCIMA and how we can help you, send for a membership application pack or call 081 672 4251 today.

HCIMA

HOTEL CATERING & INSTITUTIONAL
MANAGEMENT ASSOCIATION

Helping you succeed.

HCIMA, 191 Trinity Road, London SW17 7HN

Please send me further details about HCIMA membership.

NAME _____

JOB TITLE _____

ADDRESS _____

_____ POST CODE _____

DAYTIME TELEPHONE No _____

INDUSTRY SECTOR _____

Food Hygiene (Market Stalls and Delivery Vehicles) Regulations 1966
Information for Employees Regulations 1965
Race Relations Act 1976
Sale of Goods Act 1979
Sex Discrimination Act 1975
Social Securities Acts 1978 and 1979
Unfair Contract Terms Act 1977

Insurance

This is an extremely important subject that often gets overlooked. First of all, some form of insurance may be a legal necessity. If, for example, you employ staff on the premises, even if only part time, you are legally bound to take out approved policies from authorised insurers against liability for injury or disease sustained by your employees in the course of their work.

Your customers, too, need to be protected and you should most certainly take out some form of public liability insurance in case you cause food poisoning or some other kind of injury to your customers. You will also need to cover yourself in case your waitresses break anything in a client's home, the freezer breaks down or you have a fire in the kitchen. You will also need to insure your premises and your means of transport.

Checklist for insurance
Employers' liability
Public liability and third party public liability
Premises insurance and consequential loss of business
Cover for all contents of the premises and consequential loss of business
Vehicle insurance
Insurance against loss of driving licence
Legal insurance against prosecution under the relevant legislation affecting your business
Insurance of stock and consequential loss of business

If you are operating from home or using an existing privately owned car, you should check with the various insurance companies that the change of use will not affect the existing policies.

Insurance is expensive and, though it is essential in that it is far better to be safe than sorry, you do want to be sure that you are spending your money in the best possible way. Start off by going to a reputable broker. He will help you to find your way through the mass of different policies, but make the final decision yourself. Whittle down the choices to two or three and then read the policies in detail and with careful attention to the small print. Consult your accountant and solicitor; they may be able to indicate advantages or pitfalls you had not noticed. Once the decisions have been taken and you have policies on the go, you must remember to pay the premiums on the due dates and to update values periodically so that the contents insurance and the like represent current replacement values.

The principle of due diligence

One of the provisions of the Food Safety Act 1990 is the principle of due diligence. If a company can show that it took all reasonable precautions and exercised due diligence to avoid an offence under the Act it should greatly reduce the possibility of prosecution.

In practice this will mean setting up a control system and taking steps to ensure that the system is working. The system must cover all raw materials and ingredients as well as prepared food. It must also include cleanliness of premises and machinery and assurances from suppliers that they too have complied with the law.

All staff should receive detailed instructions as to their duties and responsibilities in the control system and should be trained in all aspects of hygiene. Consumer complaints must also be built into the system.

Chapter 4

Choosing Your Company Name

It is worth giving a good deal of thought to the choice of a name for your business, for once you are operating, you will be stuck with it. It takes enough time to build up a reputation without losing it and having to start again just because you no longer like the name under which you are trading.

Points to consider

The name you choose can set the whole tone for your operation. Here are some points to bear in mind when choosing a name.

Is it memorable? By and large, short names are easier to remember than long ones but, if a longer name is particularly apt, it may well stick in the mind even more.

Is it easy to pronounce? If so, it will be easy to communicate to others and word-of-mouth recommendation is extremely valuable to a catering company. Words which are difficult to pronounce do not get spoken so often for many people will avoid using a word if they are not sure of its pronunciation.

Does it clearly establish the nature of the services you are offering? Clever names are all very well but, if their message is not immediately clear, their business may be overlooked. Prospective customers leafing through the Yellow Pages or reading your advertising must be able to match your activities to their needs.

Does the name create the right sort of image for your business? There is no point in going for the wrong sort of slang expression if you are looking for upper-crust business. Equally, a very grand name will tend to discourage customers looking for something fairly cheap and cheerful! Your name should inspire confidence among your potential clientele and should convey the impression that you are a professional outfit and know what you are doing.

Does anyone else trade under a similar name? Obviously, you cannot trade under the same name as someone else – particularly if they are already operating as a limited company. This would infringe the legal protection of that company's name. Even if the other company had no protection for their name, you would hardly want to attract business for them, which could happen if you set up a particularly successful service. Check the Yellow Pages in your own and surrounding areas before going ahead. If you are setting up a limited company the name would then have to be checked with the Registrar of Companies before you could complete the formalities.

Does the name lend itself to the design of a logo? Some companies, particularly those with long names, operate with a shortened version of the name or with the initials. This kind of abbreviation can lead to the use of a monogram or logo made up of the initials. In other instances, a drawing is used in conjunction with the company name to make a letter-heading, which can help to make the business name more memorable. Maple Leaf Catering, for example, might use a drawing, realistic or stylised, of an actual maple leaf, whereas Jackson and Rhodes Ltd might use the combined letters J and R.

Will the name be acceptable to the Registrar of Companies should you wish to set up a limited company? The use of words which could mislead the public by suggesting that an enterprise is larger or more prestigious than it really is is a case in point.

Names fall largely into four categories. Examples, both real and imaginary, are given below for each category. Have a good look through them and see which type you think would be most appropriate for you.

The proprietor's name
Both forenames and surnames feature among the many catering businesses in the telephone directory. Some are very formal, such as Ring and Brymer, Blakes and Rothsey Clarke. Others are chattier with names such as Claire's Kitchen, Chez Paulette or Jane and John. Some of these stand as they are

but most of them are qualified by the words Caterers, Catering, Party Service or Picnics. This is only sensible, for otherwise potential customers will have no idea that your services meet their needs.

The location
Some catering companies such as London Cooks, City Caterers, Country Cooks and Aylesbury Catering have taken their names from their locations. This can be a good idea provided that the name does not seem to imply that the caterer will not or cannot operate outside the area designated.

The food
By far the most popular names are those which have a connection with eating or with entertaining. Gluttons, Plum Duff, Food To Your Door, Fresh-Bite, Party-Fare, Carvers, High Table, Party Planners, Above the Salt, and Travelling Gourmet are just a few of them.

General themes
Other names have been chosen with reference to all kinds of ideas, symbols and sayings, which presumably just happened to appeal to their owners: By Word of Mouth, Black Cat Caterers, Maple Leaf Caterers, Other People's Houses and Contessa. There may be no link between this kind of name and the services offered and so here again you may consider adding a word such as 'Caterers' or 'Cooks'.

Once chosen, the name will be used on all stationery including letterheadings, invoices, cards, compliment slips, menus and price-lists and also on your premises, delivery boxes, labels and perhaps your van or car.

Remember that if the name you have chosen is different from your own, then the Companies Act requires you to show the latter on all stationery and at your place of business, and this applies whether or not you are organised as a limited company.

Your name and the stationery on which it appears will be an important ambassador for your business and it is worth paying a good designer to suggest possible logos, together with layout and lettering. Choose just one design and use it on everything. You will achieve much more impact from constant repetition than you would by using a different layout or lettering on each item. You may also decide to choose a house colour at this stage but do not just follow your own personal preferences. Choosing red because you like it may not be as sound as choosing royal blue because it projects a solid professional image. Your designer should also be helpful here.

Remember that the end result is supposed to help you to project an image of your business. Go back and look at your business plan. What sort of image do you think will appeal most to your potential customers? Try to define the kind of image you want to put across and then explain this to the designer, who needs to be thoroughly briefed if he or she is to produce a design that really works for you.

Chapter 5
Location and Premises

For some businesses location is vital. For a catering business location is important but it will not have the same impact as it might on a restaurant or a wine bar. This is just as well, for many catering businesses are set up from the proprietor's own home. If you are starting off from home, the location is fixed and you can only re-check your market research and plan to specialise in those areas around you which seem to offer the most profit. The advantage that an outside catering business has over a bar or restaurant is that it is mobile and can take its services into those areas that have more potential custom.

Location

Of course, if you have decided to buy or rent premises especially for the business or you are planning to run a home-made delicatessen shop in conjunction with your catering service, location can become more critical. Nearness to potential customers will be particularly important to a shop, but an area with low rents may be more important for premises which will only be concerned with preparation, cooking and storage.

Being near potential customers may involve proximity to the business section of the town or being in a wealthy residential area. Whatever your speciality, the majority of your business will come from word-of-mouth recommendation. If you are planning a shop, then nearness to potential customers will also mean good passing trade, and it may be that the High Street will be a far better location for it than a small block of surburban shops.

The only way to check on passing trade in any location is quite literally to stand and count it. Do this at potentially quiet times as well as during peak shopping hours. Have a look, too, at the other shops in the immediate area. You can learn quite a lot about the location and indeed about the inhabitants of the town itself from the shops in the High Street. The presence of McDonald's, for example, usually means plenty of passing trade, and a Marks & Spencer shows a certain spending power. Check the quality of the shops in the neighbourhood and their merchandise. Do the greengrocers sell out-of-season fruit and vegetables along with the cheaper ones, what sort of specialist lines do the grocers carry, are there any high fashion names on show in the boutiques or dress shops? The answers to such questions will point to the sort of area it is.

Some areas such as Enterprise Zones are particularly attractive as they are almost free of planning controls and exempt from rates. Industrial and commercial premises in Assisted Areas are available on flexible terms and Scottish Enterprise can also provide a range of sites. The Rural Development Commission in England also provides small units directly or in partnership with the local authority.

Buying or renting premises

The decision on whether to buy or rent depends not only upon locational considerations but also on the requirements of the business, the amount of money you may be able to raise, and what is available. Unfortunately, the last factor often exerts more influence than the others. Time is usually short and, unless you are already operating from home and simply want to expand, or you have a permanent job to keep you going until you have found an approximation to the ideal premises, the pressures will be on to find somewhere as quickly as possible.

Check back to your business plan and then write out a specification for yourself which you can refer to as you inspect properties. Here are some of the points which might need to be considered when checking a prospective property.

Nearness to potential customers. See the remarks on location above.

Previous use. Has the building been used for similar purposes? If not, you will almost certainly have to get planning permission. You will also need to check the facilities with more care to see that there are no problems with essential services. The fact that there is a tap or drainage pipe in place does not mean that it is attached to anything!

45

Approximate size and layout. Will the layout affect the way in which you can use the premises? Do you need or want to be all on one floor or on different floors? A catering operation with function rooms, for example, may be much more convenient to run if it is all on one floor, whereas a shop and catering business could operate quite well with the kitchens on one floor and the shop on another.

Structural changes. How much will have to be done to the building to make it suitable for your use and will it take the structural changes involved? And, of course, the next question is, can you afford to do all that needs doing? Check the layout of water pipes and mains services against your plans for they could affect the arrangement of your kitchen.

Approximate cost range, including the desirability of freehold, leasehold or rental. If you have decided on rented accommodation you will need to look at the lease carefully to see that there are no prohibitions of use and that there is an acceptable renewal clause.

When you have a property which seems to match up to your specifications, or at least approximates to them, you should bring in your professional advisers. Your solicitor will check the lease and any other legal factors and your accountant can advise on the price level and costings. To this battery of advice you should also add a surveyor. His report could help you to avoid numerous pitfalls.

Once you have the benefit of all this professional advice, stop and look again at the premises in the light of your business plan and specification. Take a long hard look. Is it really the property for you? Do not settle for the first premises you see, however good they appear. Inspect half a dozen or so to get a feel for what is available. Finally, give some thought to why the previous tenants are moving out. Do not just accept the reasons which are given to you. Dig a little deeper. There may be facts unknown to you and which may affect your decision to buy.

Refitting

Whether you are looking at the refurbishing of your own kitchen at home or you are starting from scratch in new premises, the first step is to familiarise yourself with the requirements of the Food Hygiene (General) Regulations 1970 and Amendment Regulations 1990 (see also page 34). You must ensure, for example, that the whole area is easy to clean and that there are no nooks and crannies which might house vermin.

You may need to install window or ceiling air extractors and ventilators, or you may need another sink unit – the regulations state that unprepared food should be washed in a different sink from that used for washing up – and wash basins for yourself and your staff. Storage of food between cooking and cooling is important, and longer term storage will also need to be looked at. Here is a checklist of some of the factors:

Kitchen checklist:

Ceiling. Easily cleanable.

Walls and woodwork. Easily cleanable with no panelling or hidden interstices at the back of fitted equipment which could collect dust or harbour vermin.

Floor. Easily cleanable and made of impervious durable material. You must almost literally be able to 'eat off the floor' and you must keep it that way.

Ventilation. Window extractor/ventilator and canopy over the cookers. Check siting of fridge and freezer equipment which can give off heat.

Lighting. The lighting should reach a minimum of 400 lux at working level. Consider whether an extra window would be preferable to a higher level of electric lighting.

Electrics. Are there sufficient sockets for all the equipment you expect to install? Have any danger points been checked out for possible overloading?

Work space. There must be sufficient space for everyone likely to be working in the kitchen and for any particularly extensive requirements such as filling picnic or luncheon boxes. The surfaces should be impervious and washable.

Washing and drainage facilities. Double sink units and separate wash hand basins for staff. If appliances such as refuse disposal units or potato peelers are to be plumbed in, are the pipes large enough to clear properly? Separate sanitary accommodation must be supplied.

Equipment. This should all be movable for cleaning and easily cleanable in itself. (See Chapter 6.)

Storage. Adequate ventilation in the larder and sufficient fridge and freezer space. Certain foods must be kept at controlled temperatures during preparation and storage. The following foods must be kept cold at a temperature no higher than 8°C or if they are already cooked and waiting to be eaten hot, they must be kept at a temperature of at least 63°C: smoked or cured fish, ripened soft cheeses, prepared vegetable salads, uncooked or partly cooked pastry, prepared sandwiches, cream cakes, some dairy-based desserts and cooked products containing meat, fish, eggs, soft or hard cheese, cereals, pulses and vegetables. From 1993 many of these foods will have to be kept at temperatures no higher than 5°C. Some Environmental Health departments offer guidelines for achieving these targets.

The storage of refuse is also very important. Do you have adequate space and containers both inside the kitchen for short-term storage and outside for longer term storage? (See Chapter 9.)

If you are planning to set up a freezer service you may need a speedier method of chilling food than simply leaving it to cool.

This is also the time to think about kitchen planning. The siting of all your equipment in relation to each other and to your work surface can be very important. Think, too, about the work flow within the kitchen area. In what sequence are you most likely to be using equipment and how many people will probably be working in each area at any one time? Draw up a sample plan and then mark on it the possible activity schedules. This will show if you have any potential bottlenecks or if a particular piece of equipment is to be in an inefficient location.

If you are planning to do a major operation with a delicatessen shop, or catering premises for hire, then you will also need to think in some detail about the fitting out of these areas. Here's an outline checklist of points to consider:

Checklist for refitting the rest of your premises
Customer capacity. How much of the total area is or should

be designated for customer use? How many people are you aiming to seat in function rooms or how many people would you hope to let browse in the shop?

Work flow. It is important to determine the flow of food from the kitchen to the function rooms or to the shop. The flow of staff going about their business and the flow of customers into and around your premises also need to be considered. Passage to and from storage areas should not be forgotten here.

Display. This is obviously a vital factor in the design of a shop and should take in the window area, shelf and wall display and the counter.

Atmosphere. After the practicalities have been considered the next stage is to think about the effect you want to achieve. Once this has been decided it will help you to choose colour schemes and accessories.

The exterior. The exterior of your premises can also be an advertisement for your business, so you do not want to run out of money before you get around to considering the outside. Remember, too, that you may have to make provision outside for delivery trucks and possibly even customer parking.

Once your premises are complete, they must be registered with the Environmental Health Department in your area and you cannot start trading until they have been passed as fit.

Chapter 6
Equipment

The range of equipment available is so enormous that it is impossible to discuss the subject in detail in one chapter. Innovations in techniques and design are appearing all the time, and any analysis tends to become out of date in no time at all.

The small caterer will naturally use the cookers and gadgets to hand, gradually buying larger, heavier duty items as business expands. Others need to equip their kitchens from scratch. In both cases the golden rule is to research thoroughly before buying. Equipping a kitchen is a major outlay whether this is done over a period of time or in one fell swoop, and as a caterer you must be as cost conscious over this as over every other aspect of the business.

Criteria for buying

Quality and durability are vitally important. The equipment will be in constant heavy use and, though it may seem more expensive at first glance, equipment that is tailored to the catering trade is built to stand up to these strains and will be cheaper in the long run. Domestic equipment (particularly processors and mixers) is not built for heavy duty work and will need replacing after a surprisingly short period.

Hygiene is another point to consider when purchasing equipment. Strict rules are laid down for hygiene in commercial kitchens (see pages 34-5). Make sure that any items you consider buying have the scratch-free, easy cleaning surfaces that the health regulations demand.

Think, too, about how easy it will be to keep clean. How easily does each piece dismantle, how easy will it be to reach the interior surfaces to clean them, do the cookers have a self-cleaning programme, how easily and quickly will the fridges and freezers defrost? You want equipment that takes the minimum amount of time and effort to keep clean, yet enables you to keep the highest standards of hygiene. Equipment that is easy to clean will keep your overheads down.

Safety in the kitchen is as important as hygiene, so bear this in mind each time you look at a piece of equipment. Do the doors open in a way which increases the likelihood of the staff burning themselves? Are the lids of the water heaters hinged or loose – steam can cause terrible scalds? What safety devices are there on the deep fat fryer? Does the meat slicer have a guard? Are the handles of pans sufficiently robust? Are the pans portable or will they be too heavy to move when full? Are the oven heights right or will there be accidents lifting heavy pots in and out?

If you have any doubts about the safety of an item, raise it with the manufacturer or check with your local Health and Safety Officer. Also make sure that your safety precautions include a first aid kit and fire extinguishers. The Health and Safety Officer will insist on these and again will advise on the most suitable types for your needs. He will also advise on where they will best be placed in the kitchen.

The efficiency of each item is a point to look out for too. You will want to know the capacity of each item, how long it takes to heat up or cool down, what the power consumption is. These details can vary enormously from one product to another and it is only by researching and comparing the different performances that you will be able to identify which is the most efficient and cost effective for your needs.

But how do you find out about the products which are available? Trade papers such as *Caterer & Hotelkeeper* are excellent sources of information as the editorial pages feature reviews of new products in both equipment and food, and they are crammed with advertisements for heavy and light duty equipment. There is also a biennial exhibition held in London each January called Hotelympia. This is well worth a visit as all the major manufacturers attend and you will have a chance to see their products in action. There are also a number of small exhibitions held in the provinces throughout the year.

Keep a look out for local sales of caterers' equipment. Here you can pick up stock from manufacturers that have gone bankrupt or equipment from caterers that have closed down. They are usually run on the same basis as auction sales with a preview day so that the public can browse through the lots in comfort and decide what to bid for.

And finally, do remember to enquire about discounts if you are buying large stocks of equipment or spending a lot of money with the supplier. Just as with food and drink, equipment suppliers will have discounts for bulk purchases or large orders. There might also be a case for buying on HP or even for hiring, so talk to your accountant.

Heavy equipment

Ovens
Used for roasting, braising and baking, ovens should have a good even temperature distribution. A poorly insulated oven will lose heat and increase your fuel bills, so check this too. Test the shelves, fittings and burners to see how easily they can be cleaned. Does the oven have a self-cleaning programme? Think about which way you want the oven doors to open and check that the height is suitable for lifting heavy weights in and out.

Grills
These can either be incorporated in the oven unit or they can come as separate pieces of equipment altogether. The sort that have doors double as ovens and can be a boon if you suddenly need extra oven space.

Frying
Both shallow fryers and deep fat fryers create a lot of grease and need to be convenient to clean. Ideally, the kitchen should have three fryers, one each for fish, meat and vegetables. The small or freelance caterer is unlikely to be able to afford the money or the space for this number and will manage by changing the cooking oil frequently. Everyone knows about the fire risks attached to frying. Look for equipment which has the very best and most up-to-date safety controls.

Boiling, stewing and steaming
This is equipment which cooks food in moist conditions, not dry. One problem with these cooking methods is the volume and weight of the water. One gallon of water weighs 10 lb so it will just not be possible to move large full pans around the kitchen. For large-scale catering, the units should be fixed and have taps for filling and draining off. If the units are free

standing, check that the height will be convenient and safe for stirring. Steaming can be done in a double boiler, a pressure cooker or a steam oven. If you are looking at pressure cookers, remember to check the safety controls.

Hot cupboards
These units are absolutely invaluable for keeping food and dishes at the right temperature over a period of time. Sometimes food has to travel a long distance from the kitchen to the dining-room. Other times, because of the quantity of food being prepared, some dishes will have to be cooked well in advance and kept warm. In both instances, the hot cupboard does the job admirably. A bain marie is another excellent piece of equipment for keeping food at the right temperature without spoiling. Both come in fixed unit and mobile trolley models.

Water boilers
These come in a wide range of sizes but when full, even the smallest will be difficult to move. With fixed units, look out for efficient in/out taps. All units should have a glass gauge which indicates how full the boiler is. They should be thermostatically controlled too, otherwise your water and fuel bills will simply boil away uncontrolled. To compare the performance of different models, check the capacities, the length of time taken to reach boiling point and the fuel consumption. Some boilers have an automatic refill system which tops up in small doses so that the water never goes off the boil for more than a few moments. Others are instantaneous heaters; the advantage here is that the water doesn't stand boiling for long periods and go flat. Be careful to check the output of this type of heater as some may be too low to be practicable.

Water heaters
The catering kitchen uses a tremendous amount of hot water each day for washing up, cleaning and cooking. Make sure the system is really efficient and is not going to let you down by running cold at regular intervals. Check the capacity, output and costs of various models before choosing.

Refrigerators
Ideally meat, fish and other food should be stored in separate

fridges but, for small firms, this is just not practicable. Careful storage can prevent pungent food affecting others, but never store raw meat alongside cooked meat for fear of bacteria spreading.

In a busy kitchen the fridge door will be opened several times an hour. Each time it is opened the temperature rises and in no time at all it will be absolutely perfect for breeding bacteria. Some catering fridges have an instant cooling system which comes into operation every time the door is opened and keeps the interior temperature at the desired level. Large companies will have walk-in fridges and cold rooms. Apart from checking the efficiency of different makes, check that the doors open from the inside as well as the outside. And with all the refrigerating units, check the defrosting and cleaning systems as well as the capacity and fuel consumption. Always keep a thermometer in the fridge and check that the temperature remains below 5°.

Freezers

The choice between chest and upright freezers will probably depend on the amount of space you have available and on the volume of your business. Upright freezers take up less floor space and are thus more easily housed where space is at a premium, but chest freezers can be packed more efficiently and will hold more. They hold large packs of food better than upright freezers as there are no shelves to contend with.

For either type of freezer you will need to keep notes on the contents: what dishes it contains, where these will be found and when they went in. Whether your turnover is large or small, the contents must be consumed in date order, otherwise the food in the most inaccessible parts will never be used and eventually have to be thrown away. Check that the correct temperature is being maintained on a regular basis.

Others

Sinks - ideally with waste disposal units, but certainly with the hygienic surfaces required by the health regulations, and lots of draining space. *Dishwashers* to cope with all the kitchen washing up and also all the crockery, cutlery and glassware that comes back dirty from each function. If you are not going to use a laundry service, invest in a good, large *washing machine, a tumble drier, a steam iron* and *ironing board.* The amount of laundry generated by catering is quite staggering - table-cloths, napkins, drying-up cloths, aprons, overalls, towels and oven gloves - all need regular laundering. It is also worth investing in a good *scrubbing and polishing machine* for the kitchen floor.

Glasses, crockery and cutlery

Buying these items can be an expensive exercise, so to start with you may decide to hire as and when you need them. When a pattern of your crockery and cutlery requirements begins to emerge, you will be able to assess whether it is worth investing in your own stocks. The next question you will be faced with is how much crockery and cutlery to buy. If the bulk of your work is small functions for up to 60 people with the occasional large reception for 150 to 200 people, you obviously won't want to tie your capital up by buying 18 or 20 dozen of everything. Better to buy six dozen sets and then hire in for the large functions.

When you do buy, make sure that you choose a standard line that can be replaced or added to. Choose designs that are plain as fussy patterns and loud colours will detract from the look of the food and may not be to your clients' liking. Also be sure to choose crockery that is durable, as it is, hopefully, going to be in constant use. Crockery and cutlery are costly items to stock up on, so it may be tempting to try to save money and buy a cheaper line. This is entirely false economy, however, as you will find it chips and breaks far too easily and you will forever be buying replacements.

Transport

To start with, you may be able to get away with using an ordinary car and hire vans for the larger functions. When your business has grown and a regular pattern of large orders is established you will want to think about buying your own van. Whether you go for the small, bread vans or larger traditional vans will depend on your work profile. If your work is mainly catering for smaller functions, you will go for the small vans. If you specialise in mass catering you will need the large ones. Shop around before buying either as you may well get some discounts and do check the cost of purchasing and running the vehicles against the cost of hiring. You want to be sure that you will make savings, and it will not necessarily be cheaper to buy and maintain your own transport when you bear in mind the cost of insurance, tax, depreciation, maintenance and fuel. You will also need to think about temperature control for the food during transport. Long distances may be a problem.

Chapter 7
Setting up the Business Systems

Accounts and records are a vital part of any business and it is important to get them right from the start. Accounts are, of course, a legal requirement for any business which is registered for VAT, and also for limited companies. You will have to satisfy the demands of the Inland Revenue too, and this is much easier if you have proper accounts.

If you are not sure just what sort of accounting systems to use, read the rest of this section and then talk to your accountant. He will be able to advise on the best methods to suit your particular type of business but, unless you are prepared to spend a fortune, do not expect him to do them for you. However, you should be able to justify the services of a part-time bookkeeper.

All too often business failures are due to inadequate record keeping. The owner feels that time spent running the business is more important than writing up the ledgers. This attitude is very short sighted for it means that a mass of paperwork will accumulate which could take weeks to sort out, whereas time spent after each job writing up the accounts and records need not be too onerous.

In addition to the time saved, the accounts and records will be much more accurate and you will be able to see trends in the business much more quickly. The kind of records to keep will vary depending on the type of business you are running, but they could include individual job or customer records which will show how accurate your estimating is, which dishes or menus are the most popular and when to start changing the food on offer. Stock control records could be important if you are offering your own label products or have a system of bulk buying for wine, groceries, accessories or accompaniments. You will also need to keep a note of the whereabouts of equipment, cutlery, crockery and glasses, and a staff book is useful if you are using a lot of casual labour.

Accounting records

The essential books are cash books. They make up a complete record of all the money coming into the business and all the money going out. The records covering money going out will probably need to be broken down into three books: the cheque payments book, the cash payments book, and petty cash. The reason for the extra cash payments book is that in a catering operation quite large sums are likely to be paid out in cash both for food and staff and these items are not really petty cash items.

Cheque payments book

This will record all the money going out of your bank account. Think very carefully about all the expenses you are likely to incur and buy the widest cash analysis book you can find. Arrange those categories across the top with three columns at one end for VAT, the total amount spent less VAT and the total plus VAT. List outgoings in date order and total every month or every three months to coincide with your VAT returns. These records will show you how much VAT to claim back, where all your money is going, and whether there are any areas which need looking at in more detail.

The category headings should include those in the sample entry shown, plus Clerical, Telephone, Gas, Electricity, Depreciation, Rates contribution, Accountancy, Miscellaneous, and perhaps others too. The sample is based on a business run from home. If you have separate premises, additional entries will be required for rent and rates, maintenance and replacement of display equipment.

It is a good idea to work out a numbering system which relates to invoices, vouchers, receipts and cheque stubs. Do not just rely on the cheque number, but number each transaction in the book and repeat that number on the relevant paperwork. Where possible it is also useful to relate the entry to the particular job. Use the job numbers from your customer records book (see pages 66–7).

Cash payments book

This will be similar in some ways to the cheque payments book but the headings are not likely to be anything like so extensive. The totals here will account for money entered in the cheque payments book to cash. Remember to include a column for cash in.

Petty cash book
This is a record of all the small cash payments made for postage, petrol, small items of stationery and the like. You will need to keep a cash float for such immediate purchases. Keep bills wherever possible and remember to include columns for cash in and for VAT. If your business is not very large this book could be combined with the cash payments book.

Receipt book
This is the income side of the cash books and will record all the money, whether in cash or cheques, that is paid into your business. It can serve as a useful guide to where the majority of your income comes from, so break down the sections to give you as much information as possible. The VAT due on each transaction must also be recorded along with a bank column in which are listed payments into the bank from your paying-in book.

The sample shown on pages 64-65 is based on a small catering company operating from home.

Invoice records

A home sales day book can be useful in keeping track of money owed to you. If you keep all your invoices in date order and you are mainly dealing with private clients who pay on the spot, you may not need to keep a home sales day book. If, on the other hand, your customers are likely to take their time in paying your bills, this kind of record will serve as a reminder to chivy them up a bit.

Use prenumbered invoices or, if you have your own printed invoice forms, remember to have the appropriate number typed in each time an invoice is prepared. Keep a copy of every invoice in a loose leaf file as a double check.

Enter every invoice in the home sales day book as you send it out or present it to a client. List invoices in number and date order and make room for information on the recipient, the total amount due, the VAT amount, and any special terms such as discounts for early payment and dates due. Finally, remember to include a date paid column. As long as the

Cheque Payments Book

Receipt no	Cheque no	Date paid	Details	Total £	Total less VAT £	VAT £
1	216740	1.5.92	Butcher	112.00	—	—
2	1	1.5.92	Wine merchant	600.00	495.00	105.00
—	2	2.5.92	Cash	500.00	—	—
3	3	2.5.92	J Smith (Cook)	220.00	—	—
4	4	4.5.92	Mutual Life	450.00	—	—
5	5	7.5.92	Florist	26.00	21.45	4.55
6	6	7.5.92	John Lewis	190.00	156.75	33.25
7	7	8.5.92	Catering Hire Ltd	76.00	62.70	13.30
8	8	10.5.92	Cash 'n Carry	240.00	—	—
9	9	11.5.92	Petty cash	50.00	—	—

Cheque Payments Book

| Job no | Food £ | Drink £ | Staff | | Hire charge £ | Flowers £ | Equipment £ | Insurance £ |
			Cook £	Waiting £				
1	112.00							
—		600.00						
1								
2	140.00		65.00	15.00				
—								450.00
1						26.00		
—							190.00	
1					76.00			
—	240.00							
—								

column remains empty it will remind you to send out statements and, if necessary, ring for payment.

Total the home sales day book monthly and do not leave any invoices outstanding without some sort of reminder. Do make sure that the initial invoice goes out promptly. You are more likely to be paid quickly at the time of the service than if you leave it for weeks before sending out your invoice.

Purchase records

For some businesses these records will be unnecessary, but if you are buying a lot of supplies on credit it may be useful to have a purchase day book. This records all the purchases made on credit together with details of the date, the supplier, the invoice total, VAT amount and the date of eventual settlement. It may also include reference to a purchase order or requisition and to discounts or penalties for prompt or slow payment.

It can be useful for a business to defer payment as long as possible and these records ensure that you do not defer them for too long or forget to pay altogether, both of which will inevitably get you a bad reputation. You will probably find that some suppliers press for payment fairly quickly whereas others are much slower. It is not dishonest to take advantage of the difference and pay the former first.

Wages and capital records

If you employ staff on anything other than a casual basis you will need to keep a wages book. This shows the gross earnings for each employee together with deductions for income tax, National Insurance, pension contributions, net pay and the employer's National Insurance contributions. You will also be responsible for deducting PAYE at source.

If you have made a large investment in capital equipment for your kitchen or you have fixtures and fittings in outside premises, you may need to keep a capital ledger. The method of accounting for these items is different from that for day-to-day purchases and should be discussed with your accountant.

From the records described above your accountant will be able to get together the information needed to compile a trading and profit and loss account and a balance sheet.

However, it will only be necessary to produce these if you are trading as a limited company.

Cash control and profitability

It is vital for any business to have sufficient money coming in to meet the demands of paying for goods and services, and in this instance, paying for food and casual staff. It is sometimes possible, particularly with private clients, to ask for 50 per cent of the estimated cost in advance of the event you will be catering. This will certainly cover most of the purchases which have to be made in advance and some of the staffing costs. As well as acting as a cash flow cushion, it also covers you for cancellations. However, not all your customers will agree to this kind of arrangement and you will need to keep a careful eye on your cash flow situation. Failure to look ahead can lead to an expensive overdraft or an irate bank manager and sometimes both!

You should really work out a budget forecast and a cash flow forecast before starting up in business and they will be an essential part of any proposal to gain added finance. Once you are operating, these forecasts can be revised in the light of actual performance.

Your budget should show the estimated monthly incomings, and against these are set the outgoings. This will necessitate some shrewd guesswork if you have not run this type of business before, so beware of over-optimism. Fixed costs are averaged out over the period of the forecast. At the base of the page are the net profit or loss figures and the sum total of these will give you an idea of the overall profit levels to expect. You may need to work out forecasts for periods of six months, a year and three years. Talk to your accountant for the most useful periods to look at.

The cash flow forecast, on the other hand, shows the same series of figures, but instead of averaging out the fixed costs over the period concerned they are entered at the time they are actually due. The profit and loss figures at the base now show if there are sufficient funds to cover a very large invoice. You may have a reasonably profitable company but could find yourself short of ready cash if a large bill is due to be paid and you have a large wedding and some other big jobs to finance. It is very difficult to predict ahead in the catering business and so the estimated figures at the top of the forecast should

Receipt Book

Date	Total £	Total less VAT £	VAT £	Job no	Business functions				
					Food £	Wine £	Buffet £	Formal £	Cocktail £
8.5.92	494.00	407.55	86.45	4	494.00			494.00	
10.5.92	2100.00	1732.50	367.50	2	1240.00		860.00		2100.00
11.5.92	865.00	713.63	151.37	3					
16.5.92	116.20	95.87	20.33	6	116.20	116.20			
16.5.92	176.50	145.64	30.88	5	176.50		176.50		
17.5.92	397.00	327.53	69.47	7					
19.5.92	527.50	435.19	92.31	1	100	427.50			527.50
21.5.92	650.50	536.18	113.83	8					

Cash Payments Book

Paid in £	Date	Receipt no	Details	Job no	Total £	Total less VAT £
500.00	2.5.92		From bank account			
	3.5.92		To shopping	1	224.00	
	3.5.92		To staff	1	177.00	
	4.5.92		To staff	2	49.00	
	4.5.92		To shopping	2	54.00	

Receipt Book

| Private functions | | | | | | Bank |
Food £	Wine £	Buffet £	Formal £	Cocktail £	Weddings £	£
296.40	568.60				865.00	3459.00
						292.70
397.00				397.00		397.00
386.50	264.00	650.50				

Cash Payments Book

VAT £	Food £	Drink £	Cooks £	Waiting staff £	Clerical £	Flowers £
	73.00	136.00				15.00
			135.00	42.00		
			35.00	14.00		
	18.00	36.00				

be updated as soon as any particularly large group of events seem to be coming up all together. It will most certainly affect your short-term cash flow forecast. It's a good idea here to keep a wall diary or bookings chart so you can see such a situation arising. The chart will also help to prevent you taking on more than you can handle at any one time.

Customer and job records

The records outlined so far will tell you how profitable your business is; they will also tell you where the money is coming from and where it is going to. What they can only indicate but cannot detail is whether your estimating and pricing are accurate. To do this you will need to keep detailed customer and job records.

These records should show your estimate of costs and the quotation given to the client, together with any relevant calculations. After the event the actual costs can be filled in. Make a note of major discrepancies together with an indication of the possible reasons for them. These records will be invaluable for your future pricing policy.

Such records will also be extremely useful in pinpointing popular dishes or menus and in keeping a note of bad payers or difficult personnel. A typical customer record sheet may be set out as shown opposite.

Staff record book

If you are using a lot of casual labour, it is extremely useful to have a record of who is reliable, how far they are prepared to travel, what their particular skills are and how they work with other people.

A small notebook is all that is required. Keep a page or half page for each person you work with and classify the booklet by job function, such as cooks, waitresses, waiters, barmen, butlers, washing-up staff and so on. Obviously, you will need to include their names, addresses and telephone numbers together with notes on the factors listed above. It might be useful to star the best people so that you can find them quickly.

Customer Record Sheet

Name of customer
Contact:
Secondary contacts if necessary:
Address and tel no
Notes on the venue, kitchen facilities, etc
Notes on preferences, idiosyncracies etc
Payment record

Job no	Date	Details	Menu Estimates
			Actual costs
Job no	Date	Details	Menu Estimates
			Actual costs

Equipment hire

Some catering businesses have large stocks of their own crockery and the like which can be used by clients. Even if you do not go this far you may often find that you are leaving serving dishes at a client's house. You may also leave some stuff behind by mistake. It is very easy to lose track of your equipment in these circumstances.

To be sure of always finding missing equipment, keep a book in which every loan is entered. You could even extend this to your own use, so if a particular dish is to be used at a particular event, the job number of that event is entered against that piece of equipment in the book. You will then always be able to see where the item was last used. Such a record might be pasted on the storage cupboard door or on the shelving.

Stock control

If your business is fairly small you may not be doing very much bulk buying, but a larger business may have quite large stocks of certain items in store. It will also be employing staff who may not be quite as honest as they seem. Security is a major problem for most businesses and it is unlikely that yours will be the exception. Other catering companies invest in their own label wine and this is another expensive item which needs to be accounted for.

Stock records need not be very complicated: a record of the date the stock went into store, its quantity, and when each item is removed and for what purpose will suffice. The time you really need to watch stock is during transfer and when it is at the venue. Bottles of wine and other items from a full bar have a habit of 'walking' if you do not have a tight system of personal control.

Office procedures

It is important to designate an area in your home or premises as an office, where you can work undisturbed. Here you can keep all your records, stationery stocks and office equipment such as typewriter, filing cabinets and home computer if you decide to go to this level of sophistication.

It is fairly easy to buy all your accounts and record books, but make sure you have worked out exactly what you will need in advance, otherwise you may find that there are insufficient columns or the design just does not suit your plan.

There are also a number of pre-planned office systems which you can buy. Some of these have been designed with small businesses in mind. Many are self-duplicating so that, with one line of writing, you have made entries in several sets of records or maybe written a cheque as well. But beware, the salesman can be very persuasive and you could find that you have bought a system which is not ideal for your particular operation. So shop around, talk to your accountant, and plan in detail before committing yourself.

Chapter 8
Brochure and Menu Design

There is no doubt that a carefully researched and well-planned menu and price-list will win you business, whereas a hastily prepared, ill-conceived one will be dropped straight in the waste-paper basket. Most people looking for a caterer will ask a number of firms to send in their brochures. Competition is likely to be strong, and unless the brochure has been pitched at the right level, it will be disregarded immediately. All the time and money spent on planning and printing will have been utterly wasted.

Planning

When planning menus and price-lists, return to your business plan (see Chapter 1). In practice, it is unlikely that a firm would restrict itself to one market segment entirely – it will usually choose its main market and then pitch its image to appeal to the segments immediately either side as well. You may have chosen to specialise in catering for a middle market of people who want good quality food at sensible prices, but are also prepared to cater for large low budget functions using mass produced food and also for small up-market gourmet functions. Alternatively, you may have decided to concentrate on delivering high class working lunches to businesses but will also cater for directors' lunches and cocktail parties. The brochure should reflect this flexibility in the copy and in the range of menus and prices featured.

Having identified your market, the next step is to research the service offered by your competitors. What is the quality and style of their brochures, do they include photographs and illustrations, are many different examples of menus given, what prices are they are charging, does the price per head include waitress service and hire charges, are ancillary services offered, is VAT included in the price, are discounts given for large numbers or prompt payment?

Obviously, the range of services and prices will differ from firm to firm but it should be possible to identify those that will be your direct competitors. Once you have armed yourself with all this information, you will be able to prepare a brochure that will compete favourably against rival firms. Clearly, your brochure won't win you every job you quote for. What it will do is maximise your chances of winning orders by creating the right image.

The brochure

Brochures should always be sent out with a covering letter. This makes the client feel that he is getting personal attention, and also enables you to fine tune the details of your services even more closely to each enquiry. For instance, you may realise that, although the client has asked for details of a three course sit-down meal, his budget will only stretch to a two course fork buffet. A covering letter can tactfully point the client in this direction, highlighting the various advantages to be gained by a change of plan.

The brochure is very important as it creates an image of your company and its services in the client's mind. Large firms may employ a commercial artist to do the design, in which case, it will be worth while preparing a written design brief. This will not only clarify your thoughts but will also give the designer a clear understanding of the image required and prevent him wasting time and money producing unsuitable layouts. Small firms may not be able to justify the expense of employing a professional designer; however, most printing firms will help out here with examples of other brochures they have produced and guidance on type-faces, weight of print etc.

Ideally, you should try to use a designer and printer who have been personally recommended. If you have no contacts in this field, approach three or four of each and choose the one that seems to offer the right quality of work at a sensible price. Any design work will probably be charged at a flat fee, but the print fee will depend on the number of brochures to be printed. The longer the print run, the cheaper the unit cost. Be careful not to be influenced by this into overestimating the number of brochures you need. It's no good reducing the unit cost but then having thousands of brochures sitting idle in your storeroom.

If there are to be no photographs or illustrations, the brochure can be livened up by choosing coloured paper or coloured inks for the printing. Design features like this need careful consideration though, as they can have expensive implications. Will coloured paper mean you have to buy coloured envelopes as well? These can be quite costly. Will the firm's notepaper have to be coloured too to complete the image? General stationery costs could rise dramatically.

The menu

There are two main schools of thought here: some firms (mainly the large ones) compile a range of set menus and work to these almost exclusively; others put together a few sample menus, purely to illustrate the style of their work, but make a point of planning menus individually for each function. The former system is less time-consuming, cutting out repetitive menu planning, and is particularly suitable for the mass market – sit-down meals for two or three hundred, fork buffets for large wedding receptions, cocktail snacks for gala nights. Once the menus have been drawn up, only the prices need updating.

Tailor-made menus obviously take up a lot more time but this can be costed into the prices charged. At the more exclusive end of the market, there is a definite call for this sort of personal service. Also, if you have a number of regular customers you will need to plan menus individually as you can't keep offering the same dishes week after week.

Menu planning will not present much of a problem if you restrict yourself to set menus or plain food. However, if you are offering an individual service or specialising in imaginative food, you may often find yourself running out of inspiration. Build up a good library of cookery books and refer to these regularly to prevent your ideas from going stale. This can be an expensive exercise, but by investing in two or three classic volumes and borrowing others from your local library, costs can be kept well down.

The classic cookery books will be an excellent source for traditional dishes but books on specialist food – eastern cookery, Elizabethan cookery, vegetarian cookery etc – will be very useful too. Some of the dishes may have to be adapted but by familiarising yourself with as wide a variety of cooking styles and techniques as possible you will build up a considerable repertoire that can be continually extended. Many clients will not be familiar with the names given to classic or foreign dishes and particularly sauces. What *is* Chicken Chasseur? What *is* a Bearnaise sauce exactly? Bear this point in mind when compiling your brochure; it might be worth giving brief, mouth-watering descriptions of each dish.

The secret of successful menu planning is not only a matter of selecting appetising dishes that are appropriate to the occasion, but also keeping the courses balanced. For instance, a rich and creamy main course should be preceded by a light starter and followed by a refreshing dessert. Make sure the vegetables complement the main course. If the main course is rich, serve vegetables that are plainly cooked. Roast and fried vegetables are delicious but may be too rich for the rest of the menu. Don't serve seafood as a starter if the main course is fish.

For cocktail snacks and finger buffets, aim at a balanced mixture of pastry and bread based items, fish, meat, dairy products, vegetables and fruit. Bear in mind the colours and textures of the food. Adding a touch of red or yellow will bring a plate to life. Spiced apricots will brighten up a dish of cold chicken – and taste delicious too. Coarse pâté should not be followed by a chunky casserole – far better to precede this with a smooth mousse or a crunchy vegetable dish. Avoid soup as a starter if the main course is in a sauce and the dessert in syrup.

The list is endless but paying attention to this sort of detail will enable you to plan balanced menus that will delight your clients, not send them away groaning with indigestion.

Chapter 9

Quantities, Purchasing and Storage

Quantities

If you are just starting up in catering, probably your biggest nightmare will be the question of getting quantities right. There is always the dread that there won't be enough food to go round and that guests will be jostling for the few last morsels. Even when it is obvious that enough food has been provided, the temptation to add just a little bit more to be on the safe side can be overpowering. Don't give in to this. On the safe side you might be, but your profits certainly won't. Every ounce of extra food is costing you money and reducing your profits. Also, your client will not appreciate seeing trays of untouched food which are the result of over-estimating portions. He will conclude that, had he employed a caterer who knew what he was doing, the price per head would have been considerably lower. This may not be true but it is the natural conclusion for him to draw.

There are practical problems associated with over-estimating quantities too. If you provide too much food you will be faced with the problem of having to dispose of all the leftovers at the end of the reception. A lot of function rooms insist that the caterer takes the rubbish away with him and you could find yourself repacking and carting large quantities of wasted food away with you.

So how do you cope with the problem? It won't take long to develop a sixth sense about quantities but this is hardly a professional attitude. Every caterer must have a set of basic guidelines from which to work and this becomes more important when estimating for large functions, as any wastage will be multiplied. Similarly, it will be important when your company grows and you have to hand down tasks like estimating quantities to your staff. The only way to be sure of making accurate assessments is to work from a list of tried and tested formulas.

Obviously, no formula can be applied across the board.

73

There will be cases where allowances have to be made and you should always ask yourself, once you have worked out the initial estimate, whether there are any circumstances which will involve increasing or decreasing the quantities.

- Will there be children among the guests who will eat only half portions?
- Will the guests have come from playing sport, in which case their appetites will probably be larger than normal?
- Will they be going on to dinner and perhaps resist eating too many of your cocktail snacks so as not to spoil their appetites?
- Is the menu too rich for large portions?
- Is the function being held in the open air when people often eat more?

These and other considerations could lead you to adjust the size of your portions either up or down.

However carefully you work out your estimates, sooner or later you will be faced with the problem of unexpected guests turning up and needing to be fed. It is, therefore, a good idea to allow for one or two extra people in all your estimates and to cost these into your price per head. When the number of unexpected guests is more than simply one or two, all you can do is instruct the waitresses to reduce the portion sizes. If your waitresses are good they will be able to cope with this portion control discreetly though, obviously, if there are scores of extra mouths to feed there will be a point when there just is not enough to go round.

The well-known rule that, as numbers of guests go up, the quantities of food come down, is only partly true and can be misleading. This rule really only applies to small gatherings where there will be more wastage in preparation and cooking than for large functions. Guests at a gathering of 1000 will have just as large appetites as guests at a gathering of 300 and therefore need the same size portions. It is true that there will be savings when you are catering for large numbers but these savings will not be in the quantities of food you provide but in the cost of that food. By buying in greater bulk, you will get discounts and special prices which will reduce the unit cost of each dish.

Estimating quantities for a mixed buffet will require a slightly different approach from the norm. Whereas the individual portion for a single main dish of, say, chicken may be 4 to 6 ounces (125 to 175 grams), the quantities allowed for a mixed buffet of chicken, seafood tart and ham mousse will differ. The chicken portion will be reduced to 2 ounces (50 grams) per head and the tart and the mousse calculated at one third each of the normal portion size. There is another point to bear in mind when estimating quantities for mixed buffets. Will one of the dishes be more popular than the others? This is usually the case and rather than disappoint people by running out of the favourite early on, you should try to provide a greater proportion of those dishes that you know are particularly popular.

Converting recipes
What about recipes – how can these be adapted to large quantities? The secret here is to convert all measures to weights and then multiply these up to suit the number of people you are catering for. Three level tablespoons of flour equal 1 ounce (30 ml). An egg white is equivalent to 1 fluid ounce (25 grams). So to convert a recipe for four people calling for, among other ingredients, 3 tablespoons of flour and two egg whites, to quantities for 40 people, the ingredients become 10 ounces (250 grams) flour and 20 fluid ounces (600 ml) egg whites.

When cooking in quantity, you will also need to think about the size and shape of the dishes you are using and work out how these will affect the cooking times. They will increase, but not necessarily on a pro rata basis. For instance, food in a wide, shallow pan will cook more quickly than food in a deep one. Joints of meat are affected in the same way – a long joint with a smaller diameter will cook more quickly than a shorter joint with a wide diameter. Get to know how long your large pans and ovens take to heat up so that you can switch them on in advance and not waste time waiting for them to reach the right temperature. It is surprising how long large utensils take to heat.

Surprisingly enough, your staff can affect the quantities of food you get through. If they are wasteful and careless, you will need to provide more food. Make sure they are all conscientious about wasting as little as possible. Train them to peel and prepare vegetables with the minimum wastage, to scrape out bowls thoroughly, to cut accurate portions, to handle food without damaging it. These tips may sound petty and hardly worth the effort but they will increase efficiency

Table of Quantities

Item	Per head	Per 100 portions
Meat		
On the bone	4-6 oz/125-175 g	25 lb/11½ kg
Boneless	3-4 oz/75-125 g	20 lb/9 kg
Stews and pies	2-3 oz/50-75 g	16 lb/7¼ kg
Fish		
With lots of bone	6-8 oz/175-225 g	35 lb/16 kg
Little or no bone	4 oz/125 g	25 lb/11½ kg
Vegetables		
Fresh green	4-6 oz/125-175 g	28 lb/12¾ kg
Fresh spinach	8 oz/225 g	50 lb/22¾ kg
Potatoes	4-6 oz/125-175 g	22 lb/10 kg
Root vegetables	4-6 oz/125-175 g	22 lb/10 kg
Pulses and rice	2 oz/50 g	12 lb/5½ kg
Sweets		
Pastry	1½ oz flour/38 g	10 lb/4½ kg
Sponge or suet pudding	1½ oz flour/38 g	10 lb/4½ kg
Milk puddings	3 to 1 pt/600 ml	16 qt/18 litres
Batter pancakes	5-6 to 1 pt/600 ml	10 qt/11½ litres
Yorkshire pudding	6-8 to 1 pt/600 ml	8 qt/9 litres
Custard	6-8 to 1 pt/600 ml	8 qt/9 litres
Fruit for stewing	6-8 oz/175-225 g	40-50 lb/18-22¾ kg
Fruit for pie	4-6 oz/125-175 g	25-35 lb/11½-16 kg
Dried fruit	1½-2 oz/30-50 g	10-12 lb/4½-5½ kg
Miscellaneous		
Soup	2-3 portions to 1 pt (600 ml)	16 qt (18 litres)
Gravy	8 portions to 1 pt	6 qt (6.75 litres)
Coffee	3 (breakfast) cups to 1 pt black	
	6 (after dinner) cups to 1 pt black	
Milk		16 qt (18 litres) for white coffee allowing 2 breakfast cups per head
		4 qt (4½ litres) for white coffee allowing 1 after-dinner cup per head
Tea		5/6 pts (3/3.6 litres) for tea
Bread: sandwiches	55-60 slices per 4 lb (1.8 kg) loaf	
toast	35 slices per 4 lb (1.8 kg) loaf	

and reduce wastage. Remember that, although the wastage from one meal may seem insignificant, if you multiply this over a year, it will be quite substantial.

If you apply all the basic principles described here you will soon find that the task of estimating quantities is not nearly as difficult as you had first imagined. The table of quantities given opposite will, hopefully, make this task even easier.

Purchasing

Although purchasing and storage are two different areas of activity, they are closely related and it is therefore sensible to look at them together. A shortage of space will restrict purchasing power. On the other hand, the storage facilities may be dictated by the firm's purchasing policy. It may be decided that larger premises with more storage should be found so that the company undertake greater bulk purchasing.

There are also other areas of overlap: cash flow, shelf life, seasonal buys, proximity of suppliers, transport costs, stock control, storage techniques and so on. Each must be weighed up so that well-planned and systematic purchasing and storage policies can be developed that will make the most effective and economic use of a firm's resources. For the freelance cook this will simply mean considering how best to arrange purchasing and storage on an *ad hoc* basis to suit the work in hand. At the other end of the scale, large catering firms will have strict purchasing and storage policies drawn up and managers charged with the responsibility of seeing they are implemented.

Patterns will change as a business grows but initially all shopping will probably be done as the orders come in. Gradually, as orders increase in regularity and size, it will make sense to bulk buy basics such as flour, sugar, tinned foods, cooking fats and oils. Then, when the firm enters the big league, and large orders are a regular event, all non-perishable goods will be bought in bulk and perishable goods bought deep frozen or delivered daily direct from the suppliers.

The choice of outlet is fairly wide. To start with, it is probably best to use a combination of supermarkets, cash and carries and small local shops. Supermarkets offer very competitive prices for branded goods in standard domestic package sizes. A number offer their own brands which will be a little bit cheaper than the well-known brands, but be careful to test the quality before buying in stocks. Special offers and loss leaders can also be picked up each week in the supermarkets, so it's important to keep your eyes open for these. However, supermarket fruit and vegetables are generally more expensive than those in the specialist greengrocers and quality may not be so good.

Cash and carry

Cash and carries can be excellent as they sell goods in larger catering sizes as well as domestic packs. They are also better geared to shoppers buying in quantity: their trollies are heavy duty to take large loads, pay out desks have large conveyor belts to ease the unloading and loading of trollies, and car parks are conveniently placed so shoppers merely wheel their trollies straight out of the building to the car or van.

Cash and carries run their own special offer promotions like supermarkets and, provided storage space is not a problem, large savings can be made on these. Remember, however, to compare their prices for all goods carefully with your local supermarket prices. The name 'cash and carry' implies substantial discounts, but often prices can be the same as, if not higher than, those in the High Street supermarkets. Another point to bear in mind is that cash and carries are normally sited on the outskirts of town or in the country and it may be more cost effective to buy in the High Street than to spend an hour longer making a round trip to the cash and carry.

Local shops

While their prices tend to be higher, there are occasions when it makes more sense to buy from small, local shops. For instance, if an order comes in at very short notice for a small lunch or dinner party the following day, it can be better to save time and pop round the corner to the shops rather than drive to the supermarket. Similarly, if a function falls just after a bank holiday, you will probably have to buy some of the perishable ingredients, such as cream, at the last moment and again there probably won't be time to shop anywhere else but round the corner.

Markets

Markets are good places to shop whether your business is small, medium or large. Local markets are good for small businesses, while medium and larger firms will probably find it worth while to buy from the large wholesale markets. Local markets are geared to the housewife and are open normal shopping hours though some may only open for half a day or on certain days of the week. The range of goods available differs widely - some markets sell only fruit and vegetables while others deal with dairy produce, meat and fish as well. The large wholesale markets open in the small hours and are usually closed by midday. In large cities, they may specialise in a particular type of produce, as do Billingsgate, Smithfield and Nine Elms (the new Covent Garden).

It will be important to weigh up the merits of the large discounts that can be obtained from these markets against the cost of detailing two or three members of staff to different markets at anti-social hours of the morning. Unlike supermarkets, there is no overall quality control. It is therefore vital to do a good deal of research and shop only from stalls whose quality can be relied upon. It pays dividends to build up a good relationship with the stallholders you use, letting them know that, while you are happy to buy from them exclusively, you will have no hesitation in switching should quality drop or prices rise unreasonably.

Wholesalers

An alternative to markets is to buy direct from wholesalers or, if turnover warrants it, direct from the manufacturer. In these circumstances, the supplier will normally insist on some assurance that orders will not only be large but also frequent. They will probably want some proof of your firm's creditworthiness and financial standing, too, if goods are bought on account. In turn, you will want assurance that your orders will receive prompt attention and that the delivery service is flexible enough to meet your needs. If delivery is available on a daily basis, there are obviously no problems, but if it is weekly or longer, purchasing flexibility is lost. On the plus side, orders can be placed at short notice by phone, goods will be delivered to the door and a month or so's credit is given.

Other purchasing factors

Cash flow will be a strong influence on a firm's purchasing policy. New businesses which have to lay out money in a variety of directions to get themselves going will obviously have a restricted amount of money to spend on food stocks. Gradually, as more orders come in, more funds can be allocated to bulk purchasing, thus reducing the unit cost of dishes. Established companies will be familiar with the pattern of their cash flow and should be able to use this information to their best advantage. There will, however, always be certain times of the year when cash flow is tighter than usual. For instance, August and January are quiet months in the trade and this will be reflected in reduced receipts a month or two later. It is worth reiterating the importance of establishing right from the start what payment terms to adopt. Chapter 11 deals with this in detail. Once the policy has been set, clients should be made aware of the terms and controls set up to ensure that no payments get overlooked. If accounts are not settled on time, profits start to dwindle and the client is being subsidised by the caterer. Invoices should be sent out promptly and chased up as soon as payment becomes overdue.

Shelf life is another important consideration when

planning your purchasing policy. Modern technology has increased the life of numerous products with developments in preserving, canning and freezing techniques, but even so, there is a limit to the length of time any food can be stored. Retail goods have 'use by' dates marked on them. Catering packs and goods from wholesalers may not indicate this information, in which case phone or write to the manufacturers asking what the shelf life is. Being aware of the life of foods is not only a guard against food poisoning, it is also a means of minimising stock wastage, which in turn means profit levels are more likely to be maintained.

Food that is in season for only a short while should be bought in bulk while prices are low provided, of course, that adequate storage facilities are available and that there is a demand for the goods. Soft fruits and vegetables are obvious examples, but there will also be fluctuations in meat and fish prices which are worth capitalising on. If you are not sure of the freezer life of any goods, there are numerous freezer books which give detailed information on this point.

Wherever a firm buys its supplies, it is vitally important that the person responsible for purchasing is experienced and competent in this field. If your company is to trade profitably you must be sure that it is buying the right quality goods at the right price and at the right time, otherwise your profits will disappear down the drain in wasted food and unnecessary expenditure.

Storage

The amount of space that is needed for storage will obviously be related to the size of your business and the turnover, but whether a business is large or small, storage space is an expensive commodity. It is, therefore, very important to make the most effective use of the existing space, even if this means a capital outlay to purchase additional shelving and storage units. It is also vital to check your storage costs on a regular basis. Increases in rent, rates, electricity, gas and salaries are edging the costs up all the time and if this increase is not taken into account, a firm cannot hope to get its general costings accurate nor achieve its planned profit levels.

As your business grows, your storage space will need to be increased. Again, this expansion should be carefully planned and calculated if it is to be cost effective. First, the existing space must be analysed: could more shelving units be added to increase space, if so what would this cost, are there any offices which could be turned over to storage, could purchasing patterns be changed to eliminate the need for more space, how costly would this be in staff time?

If the answers to these questions are negative, then it is likely you have no alternative but to find additional premises. If you are very lucky, there may be premises adjacent or nearby for purchase or rental, but if this is not the case, you will have to consider moving to larger premises altogether. Needless to say, this is an expensive exercise and should only be undertaken if there are strong indications of a sufficient upturn in trade to support the moving costs and the general increase in overheads.

Whether you are redesigning existing storage space or starting with new premises, the points to consider will be the same.

- What should the proportion of shelf storage be to fridge and freezer space?
- Is it more economic to buy shelving units and cupboards or to have these built by a carpenter?
- Do the surfaces of all units meet the hygiene regulations?
- Will domestic fridges and freezers be adequate or will it be necessary to buy trade units which have larger capacities and booster cooling systems for maintaining constant and precise temperatures when they are frequently being opened? Many companies offer trade discounts so remember to enquire about these and compare costs before deciding on any item.

In large store-rooms, shelving and floor stresses have to be calculated to make sure they will bear the maximum loading. There must be sufficient space between shelving and storage units for wheeling trolleys and loading and unloading goods. Fridges and freezers should be on wheels so that they can be moved easily for cleaning. Shelving and cupboards should be designed for thorough and easy cleaning. Plans will need to be made for foods that have to be stored separately. Cooked and raw meats, for example, should never be stored in the same units. Pungent foods such as fish, cheeses and herbs should be stored separately so that they do not taint other items. There must be adequate ventilation and temperature controls, possibly with booster cooling facilities for hot weather.

Stocks will need to be dated and stored systematically to ensure that the oldest is always used first. Stock control systems should be set up to make reordering a simple procedure, and a security system established to guard against pilfering.

You should also institute a pest control regime of regular cleaning. Health officers from your local authority inspect catering premises from time to time and, if they are not satisfied with hygiene standards, they will insist on alterations being made. As these can be costly, it is worth getting things right at the planning stge, so ask your local Health Officer to come and inspect your premises and advise on the standards required. By taking such precautions, and planning and reassessing your purchasing and storage policies, you will stand a far better chance of building up a sound and healthy business than if you simply leave them to develop on their own in a rather haphazard way.

Chapter 10
Staffing

A very important factor in running a successful and profitable business is getting the staffing levels right. This applies to small firms just as much as to large ones. If a business is overstaffed, substantial sums go out every week in unnecessary wages. Added to this, output per person will often be reduced as the pressure levels are too low to create the right atmosphere of vitality and enthusiasm.

A new company is in the happy position of being able to build up its staff as the business grows. Nevertheless, it must assess the staffing needs accurately at every stage of development, otherwise over- or under-staffing will occur. Established companies need to review their staffing on a regular basis to check that they are operating at the most effective level. They must consider too whether the staffing levels and staff training policies are adequate for any planned expansion of the business.

A catering company offers a wide range of services to its clients and it must therefore employ permanent or temporary staff who are experienced in all the different skills that these services demand. Initially, staff may have to perform a whole variety of jobs, helping out wherever necessary. As the volume of work increases, the company will be able to employ more staff, who can then concentrate on their own areas of work. Whichever the case, what sort of staff does a catering company employ?

Management

As management is concerned with the direction and coordination of work, it seems sensible to consider this aspect of staffing first. The term has rather grand connotations and implies high-powered decision makers who are remote from the practical problems and dramas of the kitchen floor. The sort of management under discussion here is, however, the very practical day-to-day management of the business.

If yours is a small business you will probably be the only manager. However, if you are operating with freelance cooks it may be that they will be responsible for both the managerial and the practical side of things. They could be asked to take care of the planning and execution from beginning to end, getting the brief from the client, drawing up menus and costings, typing the quotes, planning the work schedules, supervising the cooking, arranging the ancillary services and overseeing the function on the day. In these circumstances the qualities to look for both in yourself and in staff are:

- *Reliability.* Can they be left to work on their own conscientiously and not let the firm down at the last minute?
- *Attention to detail.* Do they have the patience to draw up all the detailed plans and schedules which are vital if the function is to go off without a hitch?
- *Flexibility.* Are they able to switch hats at a moment's notice, using their creative skills one minute to plan menus and the next turning to detailed schedule planning?
- *Self-motivation.* As supervision will be minimal, do they have the self-motivation and discipline to see projects through from start to finish, applying the maximum effort at every stage?

In larger companies the manager concentrates on seeing that each job passes smoothly from one stage to the next without becoming himself involved in the physical work. His job will entail developing effective systems and chains of communication so that each project passes from one department to the next with no duplication of effort and everyone being aware of what is going on.

Management of staff on this larger more formal level demands the following qualities:

- *Organisational skills.* Do they have a flair for organisation? Are they able to think clearly and logically and not get bogged down with detail?
- *Personnel skills.* Do they have a genuine interest in people? Are they able to lead and inspire their staff with enthusiasm? Can they assess training needs and develop on-the-job training systems?

- *Planning and development.* Are they likely to be able to assess accurately future staffing levels and plan for this in advance?

Office staff

The work carried out by office staff will be the same whether the company is large or small. The actual handling may differ slightly but in the main, it is only the volume which changes. The work will entail answering the telephones, taking messages, dealing with enquiries, typing menus, letters and invoices, filing, ordering stationery and handling the mail. Petty cash and book-keeping may also fall to the office staff except in large companies where it would be handled by the accounts department.

Office staff can often feel like the poor relations of the company, remote from the more glamorous side of the business. Get them involved by keeping them up to date with the way the business is growing, tell them about the more interesting and unusual jobs that come in and relate some of the funny incidents that happen at functions. They will then begin to find their work more interesting, seeing how it links in with all the other departments. They will also begin to appreciate the importance of their own jobs, carrying out conscientiously what may before have seemed unimportant tasks.

Cooking staff

The company's reputation will obviously stand or fall on the quality of the cooking, so the importance of employing good cooks cannot be over-emphasised. This is not to say that the cooks have to be gourmet or cordon bleu standard. What they must be is skilled in preparing the sort of food in which the company specialises.

All cooks, whether freelance, part time or full time, should undergo a rigorous test to see that they know what they are doing. The test should not only cover cooking techniques but also health and hygiene. If a cook does not appreciate the latter aspects, not only is there a fair chance that he or she will produce food, at some stage or another, which causes food poisoning, but also that you will be faulted for not taking due care of your customers. Smaller firms will probably only be in a position to employ ready-trained or skilled self-taught cooks. Larger companies may take in trainees who start off preparing vegetables and acting as general helpers, but are gradually schooled in all the different aspects of cooking and presentation.

Kitchen staff and store-room hands

Neither kitchen staff nor store-room hands need to be skilled as they can be trained on the job quite easily and quickly. Honesty and reliability will be the crucial factors in selecting staff for these jobs as they may well have access to food and drink of considerable value. Sometimes the jobs are filled by trainees who are keen to work their way up through the business. In other instances, they are held by people who are not ambitious and simply want a steady job which will present them with few challenges or responsibilities. In both cases, the work should be carefully supervised and checked to see that the correct standards are being maintained. In the course of their work, both kitchen staff and store-room hands may discover ways of improving existing methods and procedures. Encourage them to think about these points and pass on any ideas they have as these could save the firm money.

Drivers and porters

These tasks may fall to any member of staff if the company is small and has an 'all hands to the pumps' philosophy. But where the volume of work is great it may be necessary to employ full-time drivers and porters. Again, reliability and honesty will be important factors in selecting these staff, as they will not only be handling goods and equipment which are worth a lot of money, but they will also have strict deadlines to meet. They must be conscientious in their work particularly regarding punctuality and attendance. You cannot afford to employ anyone who is late or, even worse, fails to turn up at the last moment. Drivers and porters must also have a responsible attitude towards security procedures. They must appreciate that they are responsible for packing and delivering items with care so that no food or equipment arrives damaged or broken. Drivers must possess clean driving licences (HGV if applicable) and be capable of carrying out minor repairs and maintenance work on their vehicles.

Waiting staff

It is as important for a catering company to have good waiting staff as it is for them to employ good cooks. First, waiting staff are in the front line and clients and guests will judge the standard of the catering partly on the standard of the waiting service. They must be neat and tidy and scrupulously clean. They must possess the correct sort of clothing and be prepared to wear uniforms where necessary. Second, waiting staff who know their job and are true professionals can save the company money. They will be fast and efficient and get the work done in half the time of less skilled staff. In fact, one of the first tasks a new catering company should set itself is to find a body of first rate professional waiters, waitresses and butlers.

Waiting staff often prefer to work on a freelance basis so it will not be necessary to take on permanent staff initially. You can simply hire them on an hourly or daily basis. Remember, though, to book them as far in advance as possible, particularly around Christmas and New Year. Later on you may want to employ a nucleus of permanent waiters or waitresses and hire in extra freelance staff as and when they are needed.

How do you know how many waiters and waitresses to book for a function? This is a bit like asking, 'How long is a piece of string?' The answer will depend on a number of factors such as how long the function lasts, how many courses are to be served, what the menu is, what drinks are to be served, whether guests will arrive together or at intervals, how far the kitchens are from the function rooms, what time the rooms must be cleared by. Having said this, the following table makes a useful starting point on which to base your estimates.

Cocktails for 30	1 waitress on food 1 butler on drinks
Finger buffet for 30 with wine	1 waitress
Finger buffet for 30 with open bar	1 waitress on food 1 waitress on drink
Fork buffet for 30 with wine	2 waitresses
Sit-down meal for 30 with wine and coffee	3 waitresses

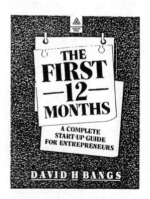

For large functions it is well worth employing a butler. He will be responsible for briefing the waiting staff, overseeing their work and ensuring that the service runs like clockwork. If there is a panic at any stage, he will spot this and pull staff off another area to help out until the crisis has passed. Clients rarely appreciate how hard waiting staff have to work, and they may think the caterer has over-estimated the number required and suggest reducing it. Don't let yourself be swayed, though. Remember that you have the experience and that too few staff will have catastrophic repercussions on the day.

Finding staff

The method you choose will depend on the type of staff you are looking for. The usual sources are:

Employment agencies
Good ones will do the initial sifting and, in theory, only send along people who are suitable for the job. They normally have a number of people on their books so should be able to fill posts quickly. Most agencies will supply temporary as well as permanent staff, which can be useful if you haven't built up your own list of freelances. The disadvantage of using an agency is that their fees can be high and, if you are hiring temporary staff, you probably won't know how efficient they are until they arrive on the day. Jobcentres offer a free job filling service and some of their offices specialise in hotel and catering jobs.

Press advertisements
This is a cheaper method of finding staff as advertising rates are usually lower than agency fees. However, it does mean that no initial sift is done and if the advertisement draws a large response, you may have to sort through a lot of paperwork and do a lot of interviewing. Advertisements can be placed either in trade magazines or in local newspapers, most of which run special sections for catering jobs. The copy deadlines for trade magazines will be longer than for local newspapers so you will have to bear this in mind, particularly if you want staff urgently. There is quite a skill in writing advertisements which will attract attention and draw the right sort of response. Study a number of advertisements before you start to draft your own, and if you feel you need

some advice, the advertisement staff on the paper will often be happy to help out.

Word of mouth

This is the cheapest way of finding staff as you will not have to pay any fees at all. Another advantage is that if you know and respect the person who makes the recommendation, then they will be unlikely to introduce you to anyone who is unsuitable. The disadvantage of this method is that it takes time and will be of no use if you need to find staff quickly.

Schools and colleges

Schools and colleges can be an excellent source of staff if you are looking for people who need training or who have only just finished their training and want to gain experience. You may become involved in a lot of sifting and interviewing, but it will only cost you time.

Training

As the Food Safety Act 1990 gets into its stride it is likely that training of staff will assume an even greater level of importance. Courses are available from many sources and can cover all aspects of the business. In the first instance you should concentrate on the training of all those who handle food. Proof of training and full records should be kept.

Contact your local TEC in England and Wales and LEC in Scotland for details of their training schemes. Training Access Points (TAPs) are available in High Street locations in many areas.

Chapter 11
Quoting for Jobs

If you are just starting up, you will probably find pricing your services a complicated and time-consuming business, but it has to be done before you can quote for a job. The prices of ingredients will not be immediately at your fingertips and you may not be quite sure of all your overhead costs. However, things do get easier as you progress and gain in experience.

Pricing your menus

The first step is to cost thoroughly and carefully every dish on your proposed menus. You will also have to do the same for any dishes which you devise specially for a specific client. Make a list of all the ingredients in a particular dish, including the tiny amounts of things like salt, pepper, herbs and spices. Take a reasonable quantity, such as that for 12, 24, 36 people or more, depending on the type of business you are planning, and put prices against each ingredient.

This will give you a basic food cost. Add to this basic cost estimates for garnishes and decorations, fuel costs for cooking, wrappings or containers and labels. Now divide by the number of portions and this will give you a cost per head for the dish. Take care with all of this for, if you underestimate prices or overlook any items, it will mean a reduction in your profit margin. This may not amount to much on individual jobs but multiplied over the year the loss of profit could be quite significant.

The next step is to cost in your overheads. This can be done dish by dish or menu by menu. Overhead costs will include some or all of the following items:

Rent and rates
Wages
Gas and electricity, excluding cooking fuel
Telephone

Transport
Advertising and publicity
Administration, stationery and postage
Bank loans and overdrafts
Clothing
Equipment
Insurance
Wastage
Laundry
Depreciation on capital equipment

Work out what the costs are likely to be for each month and remember to add in your own renumeration under wages. Divide the total by the number of people you expect to cater for during the same period. You now have a fixed cost which can be added to every per capita estimate of the food costs. This final sum will give you the minimum you must charge to cover all your costs and remain in business. There will, however, be many occasions when you will be able to charge more. The above figure tells you the lowest you can possibly go in any quotation. It does not include a percentage profit over and above a reasonable wage for yourself and this gives you some scope for juggling with prices.

When you have costed your menus, compare your prices with those quoted by your competitors to make sure you are not under- or over-charging. Double check all your estimates on a regular basis by comparing estimated against actual costs and profits. Initially, it will be worth doing this on a monthly basis so you can spot any shortfall at an early stage and make the necessary price adjustments.

Specific quotations

If you are sending out sample menus you should make sure that the potential customers know exactly what the prices include and what they do not. Usually, the extras include equipment hire, waiting staff, flowers and other ancillary services.

When it comes to producing a specific quotation for a particular job, it is important to make sure that all these items are included in the quotation or the client may refuse to pay for them when he gets the bill.

For jobs such as small buffets or desk-top luncheons, some

of these charges may look better costed into the overall price. The delivery charge for equipment hire, for example, can push up the per capita level on small orders. This does not look so off-putting if it is included in the overall cost of the meal.

Terms

Your quotation should also clearly state your terms of business. These may be different for different types of customer. For example, you may be able to get some of the money in advance from smaller customers or from private clients. With larger business clients you may have to go for payment within 28 or even 56 days. Some firms offer a discount for prompt payment. If you do this, remember to cost the discount into your charges.

It can be quite important to quote job numbers and order references, so check that these are clearly indicated on your quotation – and again on invoices – and that the name and address are correctly typed.

Chapter 12
Cocktails and Drinks Parties

Cocktails come and go. Sometimes they are in fashion; sometimes they are not. But drinks parties remain a useful way of entertaining a large number of people. This type of function can be fun for the caterer too. It offers the chance to go to town providing creative and imaginative cocktail bits. The call for simple food still exists, of course, and may prevail depending on the area. Look carefully at your catchment area to see what the demand is likely to be.

Whatever the market, it is almost certain that other caterers will be competing for the work and that the client will ask three or four companies to submit quotes. Preparing quotes is dealt with in detail in Chapter 11, but the main points to discuss with the client before drawing up menus and prices for a cocktail or drinks party are:

What is the celebration?
What time will it begin?
How long will it last?
How many guests are expected?
What age will they be?
Where will the party be held?
What floor will it be on?
Are there kitchen and cooking facilities?
What floor are these on?
Are there parking facilities?
Is the client restricted to a budget?
Is quality more important than cost?
Are there any special food requirements due to nationality or
    religion?
Will glasses and other extras be required?
Will waitress service be required?

Discussing these points with the client will give you a feel for the style of food he is looking for, the sort of price he is prepared to pay, and any additional costs you are likely to incur and will need to cover, such as equipment hire or car park fees. Unfortunately some clients think of a drinks party as a cheap option and you will need to convince them that really attractive canapés are fiddly and time-consuming to prepare and therefore relatively expensive. However, the good news for clients is that they should be able to entertain more people for a set budget than if they opted for a full meal.

For large parties it is usually safe to quote on a price per head basis but for small ones of, say, 10 to 15 people this is quite uneconomic: far better to quote a flat fee plus the cost of the food and preparation. For simple events, it may just be a case of buying in prepared foods such as quiche and pizzas, adding a few home-made items and presenting them attractively. For more exotic parties you will need to provide an interesting and imaginative selection of individually prepared and beautifully decorated items. The time of year and the facilities available may dictate whether the food should be hot or cold. Sometimes a mixture of both is a good idea.

To maintain profit margins, you may have to mix expensive and economical recipes. Consider which items will be quick and which will be time-consuming to prepare. Think of colour combinations and how the food can be presented when it is laid out on trays. Check there is a balance of meat, fish, bread, pastry and vegetables. And most important, consider how easy it will be for guests to pick up and eat each item. There are also a number of old favourites which everybody loves and asks for, such as vol au vents, sausages, and anything with smoked salmon.

Some firms send out lists from which the client selects a given number of pieces. Others prepare a selection of menus for the client to choose from - eight to ten is more than sufficient. Inevitably with the latter, clients will want to incorporate at least one item from another menu or to add some particular favourite of their own. When this happens, remember to check whether the changes have affected your profit margin - it may be necessary to increase your prices. With the list method, too, clients may choose all the expensive or fiddly items and prices may need to be adjusted.

When you are preparing your menus, bear in mind what other commitments you will have at the time. If you are likely to be short-staffed or excessively busy, include items which are quick and simple or which you will be batch cooking around that time. This sort of streamlining will enable you to offer a higher standard of food and service in the long run.

Ideas for food

Here are some ideas for drinks party canapés

Simple Canapés using Bought in Foods
Deep Fried Scampi with Tartar Sauce
Squares of Hot Pizza
Vol-au-Vents Filled with Canned Ratatouille
Falafel with Tahina Sauce
Smoked Salmon Triangles on Brown Bread
Cocktail Kebabs with Frankfurter Slices, Cubed Pineapple
and Smoked Cheese

Ever Popular Canapés
Hot Cocktail Sausages with Mustard
Croques Monsieur
Bacon Rolls with Mushrooms, Chicken Livers, or Prunes
Smoked Salmon Pinwheels
Chicken and Mushroom Vol-au-Vents
Squares of Quiche

Cold Canapés
Parma Ham Wrapped Round Banana Slices
Mixed Pâté Canapés
Mushroom Caps Filled with Liver Sausage
Blue Cheese and Walnuts in Celery Sticks or Dates
Nachos with Guacamole
Profiteroles Filled with Salmon Mousse

Hot Canapés
Spicy Meatballs with Salsa
Goujons of Sole or Calamari Rings with Tartar Sauce
Tortilla Squares
Deep Fried Cheese Puffs
Goujons of Chicken Marinated in Chinese Sauce
Hot Filo Pastry Parcels Filled with Broccoli and
Cream Cheese

Exotic Capapés
Chinese Prawn Toasts
Egg and Caviar on Cucumber Rounds
Slivers of Rare Roast Beef with Stroganoff Dip

Indonesian Satay Selection with Peanut Sauce
Chinese Grilled Dumplings
Smoked Salmon Wrapped Round Water Chestnuts
Sun Dried Tomato and Goats Cheese Tartlets

Vegetarian Canapés
Cherry Tomatoes Stuffed with Pesto Sauce
Savoury Cheese Truffles
Goujons of Smoked Tofu with Teryaki Sauce
Chicory Spears with Orange Tabbuleh
Egg and Tarragon Vol-au-Vents
Deep Fried Cauliflower Florets
Slices of French Onion Tart

Planning

Once your client has chosen his menu and confirmed the booking, it is important to start the planning immediately. Even if the function is some months away, it is worth doing some of the groundwork at this early stage. You never know how busy you will be nearer the time and, if you're dealing with large numbers as is often the case with drinks parties, you will need a longer run-in period. These initial plans may be no more than jottings on the file or action notes in your diary, but they will minimise the risk of your waking up one morning to find the function is only days away and nothing prepared.

Many cocktail items can be frozen, so if you use a freezer, draw up an advance cooking schedule. Pastry cases, vol au vents, deep fried croûtons and bonne bouche cases will freeze beautifully. Pizzas, flans, sausage rolls and sausages can be cooked well in advance, though the sausages will need a good wipe to remove any excess fat before freezing. Uncooked croque monsieur will freeze, leaving only the frying to be done on the day. Cocktail canapés are generally delicate, so it is important to pack and store them carefully.

If you are not familiar with the premises where the party is to be held, it may well be worth making a recce, preferably with the client. The size of the rooms will dictate the number of guests - allow 70–100 sq cm/2½–3 sq ft per person and the proximity of the kitchens will dictate the number of waitresses you need. Check the delivery point for the drink and food. This may be a long way from the kitchens and it is useful to know in advance so that you can bring extra staff to cope with the unloading. The client may have told you that there was masses of room and plenty of tables, but inspection of the premises may show that this is just not the case and that alternative arrangements may have to be made. Check, too, at what time you can have access to the rooms and whether they must be vacated by a certain time.

Depending on the time of year, you may need to book staff well in advance, so book them up early. Christmas and New Year are obvious examples, but there will be other periods which are affected by local events such as large race meetings, golf championships, exhibitions, conferences and civic functions. Remember, too, that the better the staff the fewer you will need. However, you should allow at least one waitress to 20-25 guests.

Equipment is not generally a problem for cocktail and drinks parties. All that is usually needed is glasses, trays, ashtrays, table-cloths, corkscrews, bottle-openers and can-openers, ice buckets, tongs, jugs, plates for the garnishes, cocktail shakers and food platters. Always provide plenty of glasses, say 50 per cent more than needed, as guests may change drinks in the course of the evening or lose track of their glasses. If hot punch is being served, check the heating facilities - it may be necessary to hire a heater.

Alcohol is a very tempting commodity and caterers have to guard against pilfering all the time. Boxes and bottles should be numbered and counted in and out by a reliable member of staff. Any drink that is being delivered in advance should be stored in a locked room or cupboard.

One of the marvellous things about most cocktail food is that it packs down small. While huge crates of food have to be carted around for large buffet parties, cocktail canapés, even for large gatherings, can be packed in a few light trays or boxes. They are, however, very fragile so will need to be packed carefully in rigid boxes. Sandwiches and bread pinwheels which have been prepared in advance will need to be well sealed in cling film - two layers may sound extravagant but it will ensure they arrive fresh and moist. Of course, the price to pay for ease of carriage is quite a long time setting up and finishing off, so allow for this.

Presentation

Presentation is very important with this kind of food. Your paté canapés may taste out of this world but they will need finishing off with a pretty garnish before they begin to look appetising. Use sprigs of fresh herbs, tiny slices of cucumber, stuffed olives, tomato and gherkin fans. Coat quarters of Scotch eggs in aspic and sprinkle cocktail sausages liberally with chopped parsley or chives. Cut sandwiches into different shapes and arrange cocktail kebabs with a view to colour and texture.

Arranging the trays of food also needs careful thought. A tray of nothing but vol-au-vents will probably lack life and colour. Instead mix five or six different items on the same tray and arrange them in diagonals or concentric rings. Place a cushion of watercress in the middle of the tray and stud them with flowers. At Christmas cover a half potato or orange with holly leaves and berries and dust with silver frosting.

The actual amount of work that has to be done on the day will depend on the menu. If most of the items have been pre-cooked and frozen, it will simply be a matter of making fillings and doing last-minute decorating. But some canapés may have to be put together just beforehand to prevent them from going soggy. Vegetable cases, such as tomatoes and cucumbers, can weep so will need to be filled at the last minute too. If the menu includes hot food it is helpful to work out a time schedule for the heating. And remember to allow a short cooling period so that guests don't burn their fingers or mouths.

You may not always be asked to provide the drink at cocktail parties, but there will certainly be occasions when you will need to offer this service. For detailed information on this subject, turn to Chapter 18.

Chapter 13

Finger Buffets

A lot of people tend to think of finger buffets simply as a cheap and cheerful way of providing a meal. More substantial than cocktail pieces but not as costly as a fork buffet, finger buffets are for people who 'can't afford to put on a proper spread'. In certain instances, this may be true, but more often than not you will find that there are very practical reasons for choosing a finger buffet. They are not the poor relation of catering, but a very flexible and useful way of serving food which holds its own alongside cocktail snacks and fork buffets in the list of catering alternatives.

Always stop and think carefully when a client asks for a particular type of food. By asking a few judicial questions, you will be able to gauge whether he has chosen the right form of presentation for the function. For example, a client asking for a fork buffet may, because of limitations of space or time, be better off with a finger buffet. On the other hand, a client asking for a finger buffet may really only require snacks or tea.

Never be afraid of questioning a client's thinking and advising on another course of action. You are, after all, the expert and many of your clients will have no experience of organising a function with food and drink. They will certainly not thank you if their guests eat only half the food because a lighter meal would have been more appropriate. Nor will they appreciate paying for a substantial menu, only to find half of it is left as the guests just hadn't the time to eat more.

Advantages of finger buffets

The time factor
You may be asked to provide lunch for 200 people attending a conference. Wine and soft drinks are to be served with the meal and coffee afterwards. From the conference agenda you

see that the delegates have only an hour for lunch. They will inevitably spend 10 minutes at each end of this hour stretching their legs, visiting the washrooms or simply unwinding, so the eating and drinking time is in fact only 40 minutes. Deduct another 10 minutes for delegates to queue for the main course and to return for the dessert and you will see that there would not be enough time to eat a fork buffet comfortably. If the food was presented as a finger buffet, however, delegates would need to make only one trip to the buffet table, and once there they would be able to help themselves to the food more quickly. Waitresses could circulate with trays of food, speeding up the proceedings even more and making the lunch a much more relaxed and enjoyable affair for the guests.

Alternatively, if the function is to be a long one starting, say, at noon and running on until midnight, guests will obviously need to be fed more than once. A sit-down meal at lunch time will keep people going until the evening, but by then they will need something else to keep them going through the evening. Having already served a sit-down meal, you will want to ring the changes and a running finger buffet is a sensible alternative.

The space factor

While some clients will choose premises to suit the size of their party, others will have to make do with rather smaller than ideal premises. A young couple, getting married and short of money, may have to hold the reception in their small flat. Businessmen wanting a working lunch may not have access to a large enough office to lay on a fork buffet. In such instances, the finger buffet provides the solution. The food can be laid out on a smaller table than would be required for a fork buffet or it can be handed round by waitresses and so avoid completely the need for large buffet tables.

The cost factor

Finger buffets do, of course, play a very important role when the client is restricted to a small budget. Finger food obviously has to be easy to pick up, and the most suitable food tends to be the cheaper pastry- and dough-based items. Prices can also be kept low because fewer waitresses will be needed, as there will be less laying out, clearing down and washing up to be done.

The food

Finger food is more substantial than canapés though some foods such as pizza and quiche can simply be cut into larger slices. A finger buffet is a meal and so there needs to be plenty of food. The best way to plan quantities is to take a dinner plate and imagine how many items would completely fill it. The answer will probably be about eight or nine.

Foreign food is often a good source of new ideas. Small versions of bruschetta, spring rolls and samosas make excellent additions to the menu, as do skewers of teryaki beef or sweet and sour prawn balls. Watch out for new ideas and you'll find you have soon collected a sizeable repertoire of menus for finger buffets.

Ideas for food

Here are some ideas for finger food:

Simple Home-made and Bought-in Finger Food
Ham Cornets Filled with Cream Cheese and Walnuts
Fingers of Spinach Quiche
Pâté and Salad on French Bread Rounds
Barbecue Chicken Wings
Roast Beef Sandwiches with Horseradish
Minted Lamb Cutlets with Grainy Mustard
Chicken Nuggets with Tomato and Basil Dip
Breaded Scampi with Tartar Sauce
Mixed Vol-au-Vents

More Exotic Cold Finger Food
Tomato Halves Filled with Spiced Rice
Egg and Anchovy Mayonnaise in Cucumber Boats
Choux Pastry Puffs Filled with Guacamole
Stuffed Vegetable Medley
Broccoli and Blue Cheese Quiche
Artichoke Hearts Filled with Ham and Peas in Mayonnaise
Italian Crostini Topped with Chicken Livers and Grapes
Green Goddess Dip, Humous and Taramasalata with Crudités
Salami Fans on Pumpernickel with Horseradish
Asparagus Rolls in Brown Bread
Mini Smoked Salmon Parcels with Crab Mousse

Hot Finger Food
Seafood Tartlets
Spring Rolls with Soy Dip
Camembert Croquettes
Spicy Chicken Drumsticks
Croques-Monsieur
Small Pitta Parcels Stuffed with Chilli
Baby Croissants Stuffed with Bacon
Curried Meatballs on a Stick
Mixed Satay Sticks with Peanut Sauce
Potato Skins with Sour Cream and Caviar

Vegetarian Finger Food
Vegetable Samosas
Cucumber Crowns with Hardboiled Eggs and Tapenade
Spiced Lentil Tartlets
Smoky Aubergine Dip with Fingers of Baked Sesame Pitta
Marinated Tofu Sticks with Teryaki Sauce
Cheese and Spinach Crispy Triangles
Stuffed Eggs Oriental
Bruschetta Squares with Grilled Aubergines and Tomatoes
Fried Cheese and Pickle Sandwiches
Ricotta and Sun Dried Tomato Quiche

You may sometimes want to offer some sweet items to finish off the meal. Ideas include:

Fingers of Home-made Treacle Tart
Profiteroles
Mini Meringues Filled with Flavoured Cream
Chocolate Soufflé Cups
Glazed Apricot Tarts
Chocolate Brownies with Fudge Topping
Pineapple, Lychees, Grapes and Orange Segments,
Marinated in Brandy and Served on Cocktail Sticks

A mixture of hot and cold items is always attractive and the buffet should take in meat and fish dishes such as herbed cutlets, breaded drumsticks and scampi as well as pastry and bread-based foods and perhaps some stuffed vegetables.

The food itself tends to be simple to prepare but it relies on clever presentation to look attractive. Arrange a mound of satay on a bed of alfalfa, breaded scallops on sliced lemons and parsley, and vol-au-vents on a bamboo tray.

Planning

Planning and organisation for finger buffets will be very much the same as for cocktail parties, though there will be a few differences. Preparation time will often be shorter than for cocktail pieces as the food is less delicate and fiddly to prepare. It is much quicker to make a number of large quiches and then cut these into slices for a finger buffet than it is to make the equivalent number of individual tartlets for cocktails.

Equipment will be the same, with the addition of small plates for guests to place their food on. More trays and meat flats will be needed for laying out the food and it is worth preparing labels, particularly if there won't be any waitresses around to explain what things are. It's not always easy to tell what sort of filling a sandwich has just by looking at it, and people do like to know what they are eating.

If the food has only to be laid out in advance and cleared away afterwards, few staff will be needed on the day. When there is a shortage of space, however, it may be a good idea to provide staff to replenish and tidy up the trays throughout the course of the party. With this system you need only put out a little food at a time, thus saving on the amount of space you have to give over to the buffet. Hot buffets will obviously entail more staff. Draw up time schedules for heating the food and remember to stagger the service; if all the hot food is laid out at once much of it will go cold before the guests get to it.

The choice of menu will dictate how much work there will be to do on the day. A lot of finger food can be prepared in advance: sandwiches with drier fillings can be made the day before and stored in the fridge overnight, well sealed in cling film. Pastry- and dough-based foods with moist fillings will have to be left to the last moment. If the food has to stand for some time before it is eaten, you will need to consider whether the room temperature will affect any of it – aspic coated foods and mousses are two classic examples of potential disaster areas. And all sandwiches will have to be left in their cling film and unwrapped at the last minute.

Chapter 14
Fork Buffets

A fork buffet can mean anything from a simple Celebration Chicken Salad followed by Trifle, to tables of traditional fare, an array of curries, a spread of vegetarian dishes or a selection of the finest *haute cuisine* dishes.

Because they are less formal than sit-down meals, there are many instances when fork buffets are the best way of serving food. For instance, a full sit-down meal may be too lavish and quite inappropriate for the occasion – perhaps when a charity is entertaining. Other times, there may be insufficient space to seat all the guests, or the host may want his guests to circulate as is often the case with the larger business functions. Time may be short, not permitting the host to seat his guests for a lengthy three course spread. Or it may be that funds simply will not stretch to a sit-down meal with all the trimmings and hire charges.

Fork buffets can be put together at quite short notice, as the food need not be fiddly, and the tableware, equipment and staffing need not be extensive or complicated to arrange. Some companies specialise in fork buffets and nothing else, taking on quite small functions as well as large weddings and business parties.

For small functions, it may be necessary to make a delivery charge or ask the client to collect the food himself as the profit margin will be small. At the other end of the scale, if the function is a really large one, the caterer will often give discounts on the price per head.

The size and formality of the function will dictate how much detail you need to obtain from the client when he asks you to quote for the job. The checklists opposite illustrate the great difference between items to check for a small and a large reception:

Christening party for 20 people

Number of guests - how many of these are children?
Any particular likes or dislikes in food?
Does client have hot or cold food in mind, or a mixture?
Discuss styles of food from plain to elaborate.
Is waitress service required?
What time will the party start?
How long will it last?
Does the client want a quote for a christening cake?
What equipment will need to be hired?
Special christening napkins or printed book matches?
Does client have a price in mind?
Is drink required - wines, sherries, spirits, champagnes, liqueurs, soft drinks, fruit juices?

Anniversary party for 150 people

Number of guests - how many of these are children?
Nationality and religion of guests?
Where is reception being held?
If indoors, what floor will it be held on?
If in a marquee, will there be electricity and water supply?
Will drink be required?
If so, will this be a full bar or just aperitifs and wines?
Is champagne required for the toasts?
Will coffee and tea be required at the end of the meal?
What time will the guests of honour arrive?
When will the buffet start?
How many speeches will there be and how long will they last?
Would the client like a quote for the cake?
Hire requirements: crockery, cutlery, glass - sherry, wine, spirits, champagne, tables, table-cloths, cakestand and knife, ice buckets.
Are table decorations and flower arrangements required?
Does client have any firm ideas for the menu?
Discuss different styles of food.
Should menu be hot, cold or a mixture of both?
Does client have a budget in mind?
What are the storage facilities on site for equipment and drink?
What is the security like?
Are there parking facilities?

Though thick steaks and all other foods that need cutting up are generally unsuitable there is, on the whole, a wide range of standard main course dishes which will be perfect for the fork buffet. Cold dishes may seem to pose a problem at first, as recipe books often don't feature many of these. However, many hot dishes such as Stroganoff or Paella are equally delicious when served cold. Try out a few and you will soon get a feeling for which ones adapt well.

The kitchen facilities on site may influence the menu planning – for instance, if there are not enough fridges on site, salmon mousse would be inappropriate on a menu for 150 people. The room temperature may also affect the choice of food, especially if it has to be laid out well in advance – will the aspic coatings melt and the salads go limp before the guests arrive?

If the numbers are large, go for dishes that are quick to serve rather than those that are fiddly to dish up or take time to carve. There is nothing more irritating than to queue up for a long time waiting to be served. With these points in mind, some sample menus are given. Several offer a selection of main courses. This is because people love to taste a little bit of a number of tempting dishes. With buffets for very small numbers it would be quite uneconomical to offer this selection.

Sample menus

Simple Fork Buffet for 20 or More
Beef in Red Wine with Caraway Rice
Seafood Pie
Buttered French Beans with Roasted Almonds and Baby Carrots
Coffee Meringue Gâteau
Fresh Fruit Salad with Curaçao

Cold Buffet for large numbers
Chicken Galantine with Spiced Apricots
Slices of Rare Roast Beef with Horseradish Sauce
Spicy Paella served with Potato and Chive Salad, Tomato and Anchovy Salad, Caesar Salad, Coleslaw
Tarte au Pommes
Hazelnut Meringue Cake

Business lunch for 20
Smoked Salmon Cornets with Prawn Mousse
Cold Beef Stroganoff served with Brown Rice
Salad, Green Salad and Avocado and Tomato Salad
Profiteroles
Coffee and Walnut Mousse

Anniversary party buffet for 150
Melon, Cucumber and Tomato Cocktail
Cold Salmon and Hollandaise Sauce served with New
Potatoes tossed in French Dressing, Cucumber and Dill
Salad, Stuffed Tomatoes and Green Salad
Apple and Orange Charlotte
Raspberry Meringue Cake

Vegetarian Fork Buffet
Gazpacho
Asparagus Mousse
Egg and Carrot Filo Pastry Pie
Courgette and Smoked Tofu Coulibiac
Lancashire Cheese Log with Sage, with Fruity Rice Salad,
Carrot Slaw and Mixed Bean Salad
Banana and Orange Charlotte
Plums in Curaçao

Advance planning

Even if the buffet is a small one you will want to prepare a work schedule as soon as the order is confirmed. In fact, it's worth doing one in your head when you are first asked to quote just to check that your order book is not too full and that you can accommodate the work. It is better to turn down the odd job rather than take on more than you can handle and run the risk of damaging your reputation with a lower standard of food and service. Turning down work is never easy, but remember that it can sometimes enhance your reputation, as clients will assume you must be good if you are that busy and the word will get round that you have to be booked well in advance.

First study your menu and see how much can be prepared in advance. Can any of it be linked with any batch cooking you will be doing for other functions? Make a note of the foods that will have to be prepared at the last moment. Some salads will fall into this category, though remember that things like pasta and rice salad can be cooked in advance, frozen and then dressed on the day. Check that there will be sufficient space in the fridges. Think about the sauces, dressings and garnishes. Which of these can be prepared in advance and which will have to be left to the last moment?

Work out a buying plan along the same lines as the work schedule. At certain times, some foods may be in short supply or in great demand, so you will want to order them well in advance; salmon around Wimbledon and Ascot, for instance, and strawberries in winter. Staff will need to be booked but, if the lead time to the function is very long, make a diary note to do this later on.

What equipment will you need to order? This will probably be no more than dinner plates and forks, dessert plates and spoons, glasses and perhaps some tableware, such as condiment sets and dishes. Specify whether the cutlery should be stainless steel or silver plate – there is a considerable difference in the price. Always order more crockery and glass than is necessary in case extra guests arrive unexpectedly or there are accidents in the kitchen. If the buffet is hot and the kitchen facilities poor, remember to order cookers, water heaters, and in really bad cases, you may even have to hire tables to work from.

If the function is a small one, the delivery arrangements will be quite simple, but if it is large these will need to be worked out carefully. Consider which articles are fragile and will need to be packed separately. Think about how heavy the crates and boxes of food will be and how many staff will be needed to load and unload these. If you haven't enough van space yourself, it may be necessary to hire some transport, whether this be a willing taxi firm or a van hire company. If any items are delivered direct, make sure they are checked off and signed for, either by someone on site who is prepared to take responsibility for them or by a member of your staff.

Book up any ancillary services such as flower arrangements, toast-masters and novelty printing for book matches or paper napkins, and make sure that all are given thorough briefs on what is required. With printed items, remember to leave enough time for proofs to be prepared and approved.

Work on the day

While the smaller functions will only require a couple of hours 'setting up' time, it will be necessary to arrive and start work much earlier for large functions. The work will be more or less the same, but there will obviously be much more of it when the guest list runs into hundreds rather than twenties. For large functions, it is vital that the day starts with a staff briefing so that everyone knows exactly what to do and no tasks are duplicated or overlooked. The main tasks will be:

Cooking or reheating the food.
Laying the buffet table.
Polishing the cutlery and glass.
Checking and polishing crockery.
Arranging food on serving dishes.
Decorating and garnishing.
Setting up the bar.
Preparing tea and coffee trays.

There will doubtless be a series of unforeseen tasks – the host may arrive with a box of oranges which he would like squeezed and served as fresh juice. Some vital piece of equipment may have been left behind and a member of staff have to be despatched to collect it or get hold of another one locally.

Although the laying up of the table will be simple it must still be done carefully. The table-cloths must all reach down to the floor and the corners should be tucked and pinned. This not only looks smart but also stops people tripping over the loose ends. All the cutlery must be polished and the crockery must be wiped and checked for cracks and chips. Forks and spoons must be neatly arranged – it's a good idea to roll them in napkins and then stack them in an attractive pattern at the end of the table, preferably the far end.

Thought should be given to traffic flow as this will dictate which end of the table guests will start from. For very large functions there may be a case for having a number of separate buffet tables dotted round the room rather than one long table. This will help speed up the serving and avoid a build-up of people waiting to be served. Alternatively, use both sides of the buffet table or make two access points on one side.

The bar too should be positioned to help the traffic flow around the room. Glasses should be arranged in an attractive pattern and the drinks laid out for ease of service. Check that all the bar tools have been delivered: corkscrews, can- and bottle-openers, sharp knife, swizzle stick, cocktail shaker, ice and ice-buckets, white cloths, lemon and orange slices, cherries, olives and cocktail sticks. Spare glasses and drink can be kept under the table so the staff don't have to go rushing off to restock half-way through the party. Remember, too, to put a few large rubbish boxes or bags under the bar and buffet tables.

If the menu is hot there will be a lot of activity in the kitchens. Test all the equipment on arrival to see that it works and remember to allow time for the heaters and cookers to heat up. Large water heaters, for instance, can take up to an hour to come to the boil and if these haven't been switched on in time, the coffee could be a disaster. Check all the food and arrange it systematically on the work surfaces to minimise the amount of running around that has to be done. See that everyone is familiar with the cooking and serving schedules. Tray up the coffee cups and saucers and stack these ready for use.

Chapter 15
Sit-down Meals

A sit-down meal can cover anything from an intimate dinner party for six to a formal banquet for hundreds of people, but whether the function is large or small, the demands on the caterer and his staff will be far greater than for buffets or cocktail parties. To begin with, there will be an increase in the number of courses on the menu. While buffet menus are normally two, occasionally three courses, the sit-down meal will have three, four and sometimes five courses. It is unusual to have a cold main course so at least one and often two will be hot. You will need staff who are skilled in silver service and can work quickly and quietly, often under extremely difficult conditions. The overall numbers will obviously dictate just how taxing a sit-down function will be.

Sit-down meals for up to 20 people

These will normally be business lunches or private dinner parties. Cooking on this scale has become extremely fashionable over the past few years and hence there is hot competition for work. Building up a book of regular clients can be a slow process, but if you are good, your reputation will spread and you should soon have a regular supply of work. Personal contacts are obviously a tremendous help in getting business, but if you are short of these, try writing to different companies telling them about your service and enclosing some sample menus.

For the dinner party market, have a leaflet printed with sample menus and have it delivered door to door to selected roads in your catchment area. There are agencies who will carry out this sort of work, but your local Boy Scouts organisation should also be happy to help out in return for a contribution to their funds. Advertisements in local newspapers, news sheets and magazines will be another way of spreading your name about and drumming up business. This is dealt with in Chapter 19.

Catering for very small parties, say two or four people, will be quite uneconomic unless you charge a flat fee plus the cost of the food. On a price per head basis, it would be very difficult to cover your costs for such small numbers. For larger parties you can charge on either basis, but do watch the costing carefully. It is very easy simply to charge for the food and forget about all the hours you have put in. When you are first starting up, and particularly with clients who are personal contacts or friends, it will be very tempting to lower your prices to attract business. While this might be a good idea short term, in the long run it will prove a disastrous policy. Your clients will become accustomed to your price range and will not readily accept any sudden increase.

Even with a small party, you will need to get a good brief from your client if you are to get the tone of the menu and service right. The sort of points you will want to cover are:

How many guests will there be?
Will their religion or nationality affect the menu?
Is it a business function or for pleasure?
What time will the guests arrive?
Will they be coming on from anywhere where they might have had cocktail snacks?
What time will they start to eat?
What time will they finish?
Should the food be plain or elaborate?
How many courses should there be?
Does the host want to impress?
Is the budget limited?
Have the guests been entertained recently, if so what did they eat?
Is waitress/butler service required?
Does the client have crockery, cutlery, glasses and serving dishes?
What cooking facilities are there?
Are table decorations required?

For the small caterer, equipment can be a problem if he or she is asked to provide lunch in an office where there are no kitchen facilities or stocks of china and glass. The crockery and cutlery can, of course, be hired for a fairly reasonable sum, but it becomes very expensive if you have to start hiring portable cookers and could put you out of the running when it comes to quoting against other larger caterers who have their own equipment. There are a number of portable cookers on the market which are not too expensive to buy and may be worth investing in. When they are not in use on a job, they make very useful additional ovens in the caterer's own kitchen.

A small dinner or lunch won't present any problems for a good waitress who is trained in silver service. But if for some reason you have to call on a waitress you have never used before, check that she is silver-serviced trained. A good waitress will also take pride in laying the table correctly, though you should always check this yourself before the guests arrive. Turning this rule around, a completely novice caterer will learn a tremendous amount by watching how the waitress sets the table and presents and serves the food. In fact, finding really professional waitresses is probably one of the most important and earliest tasks a caterer should set himself. If the function is in the evening, you may need to cost in evening rates of pay for staff, and travelling costs will rise if they have to go home late at night and therefore take taxis.

One of the problems with small sit-down meals is that they never start on time. With a business lunch, the morning meeting or preceding talk always seems to overrun, and with private dinner parties there is always at least one guest who gets delayed. It is, therefore, useful to bear this point in mind when you are planning menus and working out the fine tuning of the cooking schedule. Trying to keep food warm in an unfamiliar oven is not easy, so dishes which are prepared at the last moment are often the answer.

Sit-down meals for up to 100 people

The numbers here are still sufficiently low for the freelance cook to be able to tackle, yet at the same time they are large enough to make the exercise financially interesting for the bigger companies. The latter will have an advantage when it comes to competitive quoting as they will not have to buy in so many services. On the other hand, some clients prefer to pay a little more for the personal attention and quality of service which they get from a small firm.

One word of warning though: while the financial returns may be extremely tempting, the novice caterer would be well advised to avoid this size of function until he has built up a bit of experience. Providing sit-down meals for large numbers is in a different league from catering for buffets, with more problems and pitfalls. By taking on this sort of work too early, you may well have a disaster which will seriously damage your self-confidence. Far better to help out at one or two large functions first and then, when you know the ropes, start tendering for this type of work yourself.

Much of the work for medium-scale sit-down meals will be for Christmas parties, conferences, weddings, club dinners and other local gatherings. While personal contacts may bring in a few jobs, most of your work will be generated by placing advertisements in local newspapers. If you advertise at the right time of year, the response will probably be quite large but remember, it is the conversion of enquiries into firm orders that counts. Keep up-to-date copies of your competitors' brochures and price-lists so you know what the going rates are. Another good source of finding work is the town hall. There is normally a bulletin board in the entertainments section where you can display a small poster or card.

You will, of course, need a good briefing session with the prospective client at the premises when you can cover the following points:

Number of guests.
Purpose of the function.
Will the function have a theme?
Will religion or nationality influence the menu?
What time will the guests arrive?
What drinks will be served before the meal?
Should canapés be served too?
What style of food is required?
Are there any budget restrictions?
What time will guests sit down to eat?
What time will they finish?
Will there be speeches, if so how many and for how long?
Is a toast-master required?
Will floral arrangements/table decorations/novelties or any
    other ancillary services be required?
Will there be a seating plan?
Will all the drinks be paid for by the host or will guests have to
    buy any or all of them?
What kitchen facilities and equipment are there?

Where are delivery points and lifts?
Will the function room seat the number of guests expected?

This information will not only be vital in helping you get your costing right but it will also be the basis for your planning and organisation.

Menu planning is the next stage and here you will want to consider, not only the wishes of the client, but also any constraints of space, facilities or time that you have discovered in the course of your briefing. For instance, are the heating facilities sufficient to enable you to provide two hot courses or will you have to restrict this to one? Will you have to choose dishes that can be plated up at the last minute because there is insufficient space to do this in advance? If guests are likely to be late sitting down to eat, would it be better to choose a starter that will stand well? Similarly, you should consider any constraints within your own firm that may affect your menu. Will you be cooking the food in-house or will you buy it in from a supplier?

You are now in a position to work out your costings and, needless to say, it is vitally important that you get these absolutely right – particularly, when you are catering for large numbers. Underestimating the cost of any item or service could turn your profits into losses. Check and recheck the costs of hire, food, staff, transportation, drink, flowers, incidentals and general overheads.

Finally, check that you have considered economies of scale and that you will make the same rate of profit if the numbers are reduced or increased, as often happens. For instance, if you cost out crockery and cutlery hire for 60 people and the numbers are increased by 10, you may find that your supplier only hires out in batches of 20 and you would be left paying for 10 settings that would not be used. If you are aware of this problem right at the start you can explain to the client why the hire charge for the extra 10 will be proportionally higher. If the problem dawns on you when you are invoicing the client, it is too late to do anything about recouping the money.

Large-scale sit-down meals

Functions for several hundred guests are really out of the scope of the small caterer, unless he or she acts purely as a coordinator, buying in the services of other companies. For the most part, large-scale catering of this nature is the domain of the medium and large catering companies.

Sample menus

Business Lunch for 20
(Summer)
Mixed Leaf Salad with Grilled Red Peppers and Pinenuts
Scallops with White Wine and Artichokes and Saffron Rice
Tiramisu
Coffee and Petit Fours

Business Lunch for 20
(Winter)
Chicken Liver Pâté with Brandy and Walnuts
Leg of Lamb with Garlic and Rosemary and Flageolet,
Boulangerie Potatoes
Cheeseboard
Pineapple with Kirsch

Dinner Party for eight
Avocado Salad with Cranberry or Raspberry Vinaigrette
Beef Wellington with Braised Fennel and Olivette Potatoes
Walnut Salad with Grilled Goats Cheese
Caramelised Oranges

Rotary Dinner for 100
Melon and Prawn Cocktail
Roast Beef with Yorkshire Pudding, Roast Potatoes and
Carrots
Pear Almandine Tart
Cheeseboard

Banquet for 400
Salmon Tartar with Dill Cream
Breast of Chicken with Orange Sauce, Broccoli, Green
Beans and New Potatoes
Individual Raspberry Pavlovas
Cheeseboard

Presentation here becomes more a matter of excellence and service. Dishes should look good as they are brought to the table but inevitably they start to lose some of their extreme attractiveness as they are served. Table centre-pieces, which should not be too high, and good glassware and cutlery help to set the scene. Shaded colour combinations of table linen, perhaps to match the decor of the room, personalised menus as well as written place names, can also enhance the table.

Finally, do remember that, important though good presentation can be, it will not make up for bad quality food or inadequate service. In the end it is a consistently high standard of food and service which will bring the business back.

Chapter 16
Special Events

One of the joys of catering is the tremendous variety of the work. There simply isn't the chance to get bored, and creative and artistic people will find endless scope for using their talents. Engagement parties, weddings, christenings, children's parties, picnics, barbecues, halloween parties, the list goes on and on. As no two jobs will be the same, this chapter does not analyse each type of function in detail; instead it sketches the basic elements, looks at the pitfalls you may encounter and gives a few tips and ideas.

Engagement parties

Don't be surprised if you find the conversion rate of enquiries to actual orders is low. This does not necessarily mean that your prices or menus are off the mark. If very often happens that the client gets carried away with the excitement of the engagement and thinks he will get caterers in to take care of everything. By the time the quotes arrive he has begun to realise how expensive the actual wedding is going to be, with photography, flowers, dresses, car hire etc, and so decides he will save a bit of money and get the family to provide the engagement party.

Engagement parties are usually cocktails or fork buffets. They tend to be informal events where the main idea is for the engaged couple, their family and friends, to circulate and meet each other. At some stage of the reception there may be a speech or two and some toasts. Some clients may want an engagement cake served after the toasts and you may be asked to provide this. Engagement parties are usually held in the home of the bride's or groom's parents or some other member of the family, so you may find that the space and size of the rooms dictate the kind of food that is served.

As always, the key to success is the briefing and the main

points you will need to discuss are:

Drinks	— soft and alcoholic
	champagne or sparkling wine for the toasts
Food	— cocktail snacks, finger or fork buffet
	plain or celebratory engagement cake
Staff	— waiters, waitresses, butlers
Decorations	— flower arrangements
	table decorations
	engagement napkins and book matches
Photography	— is a professional photographer required?
Timing	— start
	finish
	when will speeches be made?
Budget	— does client have a figure in mind?

The sort of menus you might then arrive at are:

Cocktail Menu
Tiny Crab Meat Patties with Dill Mayonnaise
Savoury Scones with Pizzaiola Topping
Ham and Tongue Sandwiches with Mustard
Mushrooms Stuffed with Tapenade
Melon Balls in Salami Curls on a Stick
Toasted Goats Cheese on Pumpernickel
Lamb Kofkas with Mint Yoghurt Dip

Drinks
Mixed Bar
(No Speeches or Toasts)

Fork Buffet
Smoked Trout Mousse
Lightly Curried Chicken Salad with Mangoes
Roast Ham Cubes in Aspic with Parsley
Stuffed Roll of Pork with Dried Apricots and Rosemary
Selection of Four Salads
Crème Brulée
Blackcurrant Ice-cream with Blackcurrant Sauce

Drinks
Wine, Red and White
Soft Drinks
Sparkling Wine for the Toasts

Weddings

Wedding parties can range from the simplest drinks reception with cocktail snacks through more substantial finger and fork buffets to a full sit-down meal. They may last just a couple of hours or go on through the day, ending up with a dance or disco in the evening. Sometimes they may be split, with a small family wedding breakfast in the day, followed by a large reception in the evening. The caterer may simply be called on to provide the food and drink. On the other hand, the client may want everything arranged – photography, flowers, car hire and so on.

The client usually has a basic idea of what sort of reception he wants, but on closer inspection of the costs involved it may be that his plans have to be tailored down or rethought to suit the budget. It is therefore a good idea to start off by sending a variety of menus so that he realises you can provide cheaper meals as well. The following examples cover a useful range of styles and price ranges of food:

Wedding cocktail snacks
Fritto Misto with Green Mayonnaise
Choux Puffs with Walnut and Cream Cheese
Mixed Vol-au-Vents
Prawn Baskets
Curried Chicken Tartlets
Egg and Anchovy Canapés
Mini Cheese and Ham Croissants

Formal sit-down wedding breakfast
Oeufs en Cocottes
Guinea Fowl with Grilled Peppers and Sun Dried Tomatoes,
Duchess Potatoes and Broccoli
Cheeseboard
Damson Mousse

Simple fork buffet for weddings
Turkey Galantine
Honey Glazed Ham Served with four Salads
Stuffed Egg Medley
Almond Trifle
Fresh Fruit Salad and Cream

Summer wedding – buffet with seating arrangements
Poached Salmon
Cold Lamb en Croûte Persillée Served with Potato and
Chive Salad, Cucumber Mousse, Avocado and Walnut Salad,
Green Salad, Tomato and Melon Salad
Raspberries and Cream
Cheeseboard

The briefing meeting will be extremely important as there will
be many details to discuss. The following checklist is designed
for both large and small receptions, though you may find you
can delete some of the questions for simple functions:

How many guests will there be?
Where will the reception be held?
How big are the rooms?
What are the kitchen facilities like?
Should there be waitress service?
How long will the reception last?
Should it be a sit-down meal, a finger or fork buffet or cocktail
     snacks?
What sort of food does the client have in mind – simple or
     exotic, traditional or continental?
Will you be asked to provide the drink?
If so, what sort – sherries, wine, champagne, full bar, soft
     drinks and fruit juices?
Will crockery, cutlery and glassware have to be hired?
Are there sufficient tables and chairs or will these have to be
     hired?
Will you be called on to provide any ancillary services – flower
     arrangements, bouquets and buttonholes, transport, a
     toast-master, photography (still and video), wedding cake?

Once you have sorted these points out you will have a good
idea of the work load, and you should check this out with your
order book to see whether you need to call in extra staff to
help out. The majority of weddings are held from May to
September so it is quite likely you will be very busy at this time,
particularly if you specialise in weddings.

Christenings

Most people arrange the catering for christenings themselves
and caterers are called in when the family feel they really can't

cope with the new baby as well as the food, or when the party is a large one. Normally, the christening party is a morning or tea-time finger buffet with alcohol and, of course, the christening cake. Larger christening parties are usually lunch-time or evening affairs with either cocktails or a finger or fork buffet.

By its nature, the christening party guest list embraces all age groups from toddlers to grannies and this enormous range in tastes must be borne in mind when planning the menu and drinks: soft drinks for the children, tea and coffee for the older generation; novelties such as tiny chocolate cakes in the shape of hedgehogs and mice; perennial favourites like hot sausages, smoked salmon rolls and vol au vents that will appeal to everyone. Remember to find out how many children there will be as this will affect the amount of food needed and the pricing – most caterers offer reduced prices for small children.

You may have to hire equipment if the party is a large one: hot water heaters, tea and coffee cups, glasses and small plates for a finger buffet. Dinner plates and forks, dessert bowls and spoons for the fork buffet, plus crockery and glass for the beverages and drink. If there are toasts, sparkling wine or champagne will be needed with the appropriate glasses. If the cake is to be a centre-piece, a cake stand and knife too. You may also have to supply special christening napkins and small boxes for sending bits of cake to absent family and friends. It is traditional to save the top layer of the wedding cake for the first christening, but for subsequent christenings the parents may decide to ring the changes and plump for something quite different from the traditional fruit cake with royal icing – an elaborately decorated sponge cake, an ice cream gateau or a meringue cake filled with raspberries and cream.

Christening tea
Smoked Salmon Quiche
Olive and Salmon Pizza Slices
Mixed Cucumber, Egg Mayonnaise and Tomato Sandwiches
Malt Bread, Sliced and Buttered
Scones with Raspberry Jam and Cream
Butterfly Cakes
Home-made Biscuits

Mocha Sponge Cakes with Chocolate Icing
Tea, Soft Drinks and Sherry

Christening buffet
Herbed and Breaded Chicken Pieces
Scampi on Sticks
Croques-Monsieur
Bacon Savouries
Hot Sausage Rolls
Egg and Smoked Salmon Canapés
Pâté Tartlets
Ham and Asparagus Rolls
Christening Cake
Wine and Soft Drinks
Champagne for the Toast

Anniversaries

Silver and golden wedding anniversaries immediately spring
to mind, but there are in fact many more, such as pearl, ruby,
sapphire and emerald, which you may be asked to cater for.

Some anniversary parties will be organised by the couple
themselves, others will be surprise affairs arranged by the
children or close friends. There is no traditional time or
format for the celebrations though there is normally an
anniversary cake and champagne or sparkling wine for the
toasts. If the client has no fixed ideas about the sort of party
he wants, find out how much money he wants to spend, how
many guests there will be and how long the party will last.

From this information you can gauge what sort of function
would be the most appropriate and guide the client in this
direction. When the clients are elderly they may find the
organising is too much for them to cope with, and you may
be called on to arrange the function from start to finish.
Anniversary menus will be similar to those used for weddings,
but you may be asked to link the menu to a theme. Apart from
the obvious 'age' theme, there may be other aspects of the
couple's life that are fun to focus on: golf, music, sailing, DIY –
all sorts of interests that can be featured in the cake and food,
the decorations and the flowers.

Sample menus

Golden Wedding sit-down meal
Coquilles St Jacques
Roast Veal with Apricot Stuffing, Croquette Potatoes, Fried
Chicory and Peas
Lemon Parfait or Home-made Coffee Ice-cream

Silver wedding cocktail pieces
Deep Fried Garlic Mushrooms on Sticks
Devils on Horseback
Hot Camembert Crescents
Celery Boats Filled with Taramasalata
Cucumber Boats Filled with Smoked Trout Mousse
Crab Pyramids
Savoury Stuffed Dates
Bacon with Water Chestnuts
Strawberry Tartlets
Lemon Curd Bites

Children's parties

On the surface, children's parties may appear to be quite
simple affairs – they can in fact involve quite a lot of creative
thought and detailed organisation.

Children's tastes are much more sophisticated than they
were 20 years ago, with the emphasis on savoury items rather
than the sweets and jellies of yesteryear. You will obviously
have to put forward ideas for the menus but check with the
mother first what sort of food the child likes, as children have
pretty firm ideas about what they want. The choice of foods
will depend largely on the age of the children, but here are
some menus for toddlers and young children.

Sample menus

Children's birthday party
Mixed Sandwich Boat
Individual Quiches
Sausages on Sticks
Brighton French Bread Marina
Individual Fruit Jellies
Ice-cream Clowns
Steam Engine Cake

Children's Christmas party
Christmas Drumsticks
Pastry Parcels
Savoury Yule Logs
Sausage Cannon
Mince Pies
Ice-cream Snowmen
Santa Christmas Cake

Themes have changed too. Whereas the cake used to reflect traditional topics such as trains, animals and nursery rhymes, film characters, computerised toys and sports and pop stars are the things that appeal to children today. Children's Christmas parties have changed less and the traditional foods and themes still stand – yule logs, cakes in the shape of Father Christmas, Christmas stockings, Christmas trees and reindeer.

Entertainers have stood the test of time, too (probably because they take the pressure off the parents and help to preserve their sanity) so there is still a call for conjurors, Father Christmases, balloon men, story-tellers and puppet shows, any of which you may be called on to book. Then there is the endless list of novelty items that children love: hats, masks, napkins, streamers, whistles, crackers, snowballs, lucky dips, jokes, balloons and so on. You may be asked to provide these so it is useful to trace your nearest supplier.

The problem with children's parties is that the parents often don't appreciate the time and effort that goes into preparing and presenting the food and they consider the caterer's charges excessive. They feel that because children are small, the cost of the party should be low. Don't be tempted to cut your prices though, and be particularly careful to cost in all your staff time when you are preparing your estimates, otherwise you will find you are working very hard for little or no profit.

Picnics

Picnic hampers are a creative and rewarding area of catering. They are usually wanted for special occasions such as a trip to Glyndebourne or a day at the races, so both the presentation and the food can be lavish.

Caterers who specialise in the work tend to have a range of

hamper menus from which the client can choose. If you only get occasional orders for hampers, you will probably plan each menu individually. Whichever the case, you will want a clear brief from the client and the sort of things you should discuss with him are:

How many courses should there be?
Any favourite dishes to be included?
Should menu be finger or fork food?
Cutlery, crockery and glass – would client like disposable or not?
Table-cloths and napkins – would client like disposable or cloth?
Hamper – traditional basket, open basket, wooden box, cardboard box?
Colour schemes – any particular choice?
Drinks – wines, spirits, beers, liqueurs, port, sherry, soft drinks, coffee, tea, any others?

Once the menu has been settled, give some serious thought to how it will be presented within the hamper. The secret of a really attractive hamper is attention to detail – think how even the smallest and most insignificant items can be presented. For example:

- Tie the corkscrew, tin opener and swizzle sticks to inside of hamper lid with brightly coloured ribbons.
- Wrap cutlery in frilly doilies and tie with coloured ribbon like a posy of flowers
- Put salt and pepper in brightly coloured paper twists.
- Present the cheeses in a mini hamper of their own.
- Carry the colour scheme through everything – the table-cloth, napkins and crockery.
- As an alternative to bowls, serve the starter in crab or scallop shells and the dessert in scooped-out grapefruit or pineapples.
- Fix a menu card to the inside lid of the hamper.

These are just a few suggestions. Think of more so that your hampers are a delight to look at as well as to eat.

There are two golden rules that must be mentioned before looking at ideas for hamper menus. First, be absolutely certain that all food and drink is carefully packed and thoroughly sealed so that nothing gets squashed out of recognition and there are no leakages. A soggy hamper with split or damaged food will be a disaster and ruin your client's day. Second, make sure nothing is omitted at the packing stage - a hamper with fine wines is no good if you have forgotten to put in the corkscrew or glasses! Draw up a detailed packing list and check this religiously before the hamper is delivered or collected. Your checklist will probably include most of the following:

Crockery	Beers
Cutlery	Corkscrew
Glassware	Swizzle stick
Table-cloth	Can-opener
Napkins	Slices of lemon, orange;
Menu card	cherries etc
Food	Soft drinks
Salt and pepper	Tea
Mustard	Coffee
Wines	Milk and sugar

Sample menus

Cool Weather Picnics to Serve from the Car Boot
Wild Mushroom Soup
Smoked Haddock Chowder
Game Pie with Cranberry Pickle
Pitta Parcels with Garlic Lamb
Bacon and Egg Tartlets
Iced Carrot Cake

Warm Weather Picnic
Quails Eggs and Spinach Tart
Stuffed Roll of Veal
Dutch Salad, Tomato and Basil Salad and Cold Succotash
Mocha Eclairs

Elaborate Picnic for Glyndebourne
Chilled Vichyssoise
Lobster Cocktail with Walnut Bread
Smoked Ham Rafts with Asparagus
Chicken Breasts Chaudfroid
Boeuf à la Mode with Four Salads
Strawberry Meringue Baskets

Barbecues

While small barbecues are simple to cater, large ones can present problems. The main difficulty is timing the food so that all the guests can eat at roughly the same time. Nothing is worse than being able to feed only about 10 at a time. There are two solutions to this problem: to hire in sufficient barbecues and barbecue chefs to cook for all the guests simultaneously or to pre-cook the food and simply finish it off on the barbecue.

The first system will depend on the amount of space available and the size of your client's budget. Obviously, hiring extra equipment and extra staff will push the price per head up and you must, of course, have the space to put them. If the client is happy to pay, however, it does mean that the food will be tastier and more authentic than if it is merely finished off over charcoal. The pre-cooking system can be cheaper if the caterer has his own portable cookers which he can bring along. But if he has to hire them, the only saving against the first system will be on staff costs.

Another problem with barbecues in this country is the unreliability of the weather. It is fine if there is room for all the guests to go indoors in the event of rain and the barbecues can be sited under cover, or if the client is prepared to hire a marquee as a fall-back, but if not, the client should be warned that rain - and you don't need a lot of it - could ruin the evening.

Most barbecues are two courses - main and dessert - but sometimes a starter will be required too. Remember that, in addition to the barbecues, tables will be needed to lay out the salads, cheeses, desserts and, of course, the drinks. If the party is a large one you may have to hire in a number of trestle tables. Also remember to allow bigger portions per head as people always seem to have larger appetites when they are eating out of doors.

Wine cups or punches are very popular accompaniments to barbecued food. They can be flamboyant and help make the evening that little bit more special. Do provide alternative drinks though, as some people do not like them. For drink suggestions see Chapter 18.

Barbecue food offers the caterer the opportunity to demonstrate his creative flair with marinades and sauces unless, of course, the client wants plain food. Make sure that your chefs understand fully the techniques of barbecuing, especially for large numbers of people. The food must be cooked at the right height from the coals – if it is too near it will taste of charcoal, and if it is too far away it takes too long to cook. The art of cooking barbecue food well is as demanding as any requirements in the kitchen.

Some barbecuing ideas

Fresh Sardines Served with Lemon Juice
Trout Marinated in Dill and Chive Dressing and Cooked in Foil
Steaks with Anchovy Butter
Orange Glazed Spare Ribs
Hamburgers with Spicy Sauce
Boned Leg of Lamb with Vinegar and Thyme Marinade
Duck on a spit with Garlic and Lemon Marinade
Seafood Kebabs with Hollandaise Sauce
Spiced Pork Satay
Lambs' Kidneys on a skewer with Bearnaise Sauce
Bacon and Banana Kebabs

Vegetarian food

There are basically two types of vegetarian – lacto-vegetarians and vegans. Lacto-vegetarians cut meat from their diets but are happy to eat milk, cheese and eggs. Vegans eat vegetables, fruit, nuts and grain, eliminating all foods of animal origins. Some vegans will not eat sugar or honey but will eat soya milk and soya flour cheese. You must therefore find out which sort of vegetarian food your client wants.

Vegetarian food should present no problems for today's caterers. There are numerous cookery books which specialise in vegetarian dishes – both lacto and vegan – and probably as many which deal exclusively with vegetable cookery. Eastern food is another good source of ideas for vegetarian dishes and many European dishes such as pizzas, pastas and quiche are either already suitable or can be adapted for vegetarian menus. The less common ingredients for vegetarian dishes are also much easier to buy today than they used to be.

Although vegetarian food is not difficult to cook, it is important to experiment so that you become familiar with the tastes, textures and constraints of various dishes. Some vegetarian dishes take a long time to cook as they contain pulses.

Ideas are given in earlier chapters for fork and finger buffets and cocktail food but here are some more ideas for special events.

Vegetarian Dinner Party
Mushroom and Olive Crostini
Blue Cheese and Broccoli Roulade with Tomato Sauce
Mixed Grilled Vegetables with Fresh Herbs
Hazelnut Meringue Gateau

Vegetarian Wedding Breakfast
Gratin of Crêpes Stuffed with Leeks
Baked Aubergines with Goat's Cheese, Okra Tossed in Olive Oil with Bulgar Pilaff
Strawberry Romanoff

Vegetarian Picnic
Small Vegetable Samosas with Onion Bhajis and Raita
Sweet Potato Plait
Whole Lettuce Stuffed with Egg and Peas
Walnut Picnic Loaf
Fruit Tartlets

Chapter 17

Ancillary Services

Catering entails far more than simply supplying food and drink. In many ways it is similar to the entertainment business: presentation must be professional and artistic, service must impress, functions must be stage-managed, timing is crucial, settings and atmospheres must be created and points must be exaggerated for effect. A catering company must, therefore, be prepared to offer its clients all the ancillary services that together complete a successful function.

This may sound a tall order for the one-man business or small firm, but it need entail no more than simply building up a list of reliable suppliers and then hiring their services as and when they are needed. Even the very large catering firms that employ their own full-time flower arrangers and cake decorators will hire in the more specialised ancillary services from time to time.

The small caterer may feel he hasn't the time to spare to organise these extra services but he should remember that the more helpful and varied the service he offers, the better and faster his reputation will grow.

It is important to check the standard of service of all suppliers thoroughly before using them. They must be reliable, professional, efficient and competitively priced. The standard of their service will reflect on you - if it is good, your reputation will profit, if not, your reputation will suffer. Build up a good relationship with the suppliers you use frequently - there will be times when you need to call on them at very short notice and if relations are good, they will be more inclined to pull out all the stops and help out.

If you haven't had to call on a supplier for some time, make a few enquiries before booking. Changes within the company may have led to a deterioration in the standard of service of which you are quite unaware. And try to have at least two suppliers for each service so that you are not stuck if one is already booked up.

The charge for supplying these extra services can be calculated in two ways. The service can be charged out at the supplier's rate and a consultancy fee added to cover the caterer's costs and profit, or the caterer can add his own percentage to the supplier's fee when invoicing the client. If the cost of the ancillary services amounts to a considerable sum, you may need to ask the client for a deposit in advance.

Though you may start off hiring in all the ancillary services, there will come a time when it makes sense to start supplying some of these in-house. Watch the pattern of demand to see whether there is justification for expanding the business to provide the extra services, bearing in mind that you may be able to hire out the services to other caterers, not just your own clients.

The list of ancillary services is large and varied. Requests can range from something as standard as the supply of marquees to the more unusual, such as the provision of horse-drawn carriages or vintage cars. Country-based caterers may find it more difficult than their city colleagues to track down suppliers of the more unusual services, but if the client is prepared to pay, it should be possible to supply almost anything.

Equipment hire

While the larger, more established firms will have their own equipment, new or small companies may have to hire in most of this, which will tend to put them out of the running when competing against large concerns. Rather than incorporate the hire charge in with the menu price, it is a good idea to separate the quote out so that the client can see how the food and hire costs break down. If he is sufficiently impressed with the menus, he may decide to bear the expense of the hire or even to make his own arrangements, such as borrowing equipment from the office canteen.

Most hire firms stock at least two styles of china, one for formal functions and one less formal. There will be choice in plate sizes and shapes too, so the order must stipulate exactly what is required, eg, demi-tasses or full-size coffee cups, soup bowls with or without rims, large or small milk jugs, china or metal condiment sets. Similarly, there will be a choice of cutlery, silver plate or stainless steel. Clear instructions must be given for glasses: are the wine glasses to be hock glasses or Paris goblets, should they be x or y fluid ounces, should beer glasses be straight-sided ones or mugs, half pint or pint? Indicate the sizes of the table-cloths as well as the number. Should they be round, square or rectangular?

Tables come in a variety of shapes and sizes, too. You may need to draw up a floor plan to work out the right combination. Allow for the table legs when estimating how many people can be seated and give sufficient space between the tables for waitresses to move to and fro. Should the chairs be the cheaper moulded plastic ones or the more expensive upholstered variety? If posts and rails and coat-racks are required, how many of each will be needed? What heavy equipment will be needed: bains marie, cookers, water-heaters, wine coolers etc? Should these be gas or electric? Most hire firms issue comprehensive brochures, illustrating their stock. If you are familiar with it, then orders can simply be placed over the telephone and followed up with a written order. If hiring for the first time, it will be important to visit the showrooms and see all the articles for yourself. The quality may not be as you had imagined and it may be necessary to find an alternative supplier.

Marquee hire

Hiring a marquee is expensive and the price often comes as a shock to the client. In certain circumstances, however, there may be very strong advantages which outweigh the cost. For example, if a client wants to throw a large party at home but finds his house is too small, a marquee can provide the additional space. If a function is to be held in the open air, a marquee will save the day if there is a downpour. Catering facilities may need to be laid on at a sports ground where there is no pavilion and here again, a marquee would be the answer.

The range of marquees and accessories is very wide and the choice should be discussed in detail with the client well in advance. The number of guests and the ground space will usually dictate the floor space needed. If there is to be a buffet and a disco, two separate tents could be more appropriate than one. When more than one marquee is needed, the shapes and sizes of each will need to be calculated. Should they be totally enclosed or should one of the sides be open? Will awnings and covered walkways be needed? Is it advisable to order floor covering? If so, should this be coconut matting, astro turf, carpeting or wooden boarding? Will it be necessary to lay on power? Does the client want lighting? If so, what form should this take: strip lighting, spot lights, side lighting, chandeliers? If the atmosphere is to be opulent, wall covering and artificial windows and doors can be supplied. Should lavatory tents with Portaloos be laid on?

As soon as the choice has been made, the order should be placed. Reputable companies will be booked well in advance and it is important to be able to rely on both the quality of the equipment and the efficiency of the company in delivering and erecting the marquee and then clearing the site afterwards. The hire company may require a deposit and this sum should be added to the general advance paid by the client. When the order is confirmed, make sure the client is informed of the delivery and collection times. Special arrangements may have to be made for someone to be on site to let the vans in. Security passes may have to be issued if the equipment is to be delivered to business premises. Finally, check that delivery and erection are going according to plan on the day – you can't afford to arrive on site with all the food and drink, only to find the marquee has not been delivered.

Cakemaking and decorating

If a company does not employ a skilled baker on its staff, this work can easily be contracted out. There are many people who supply cakes on a freelance basis and most bakers will make and ice cakes to specification.

Choose suppliers who can cope with anything from the traditional wedding cake to novelty cakes calling for imagination and ingenuity. Put together an album with coloured pictures illustrating the range of cakes available. Offer a selection of each type of cake: semi- or fully marzipanned cakes, fruit cakes with or without alcohol, fat-free or flourless cakes, sponge cakes with cream, butter or jam fillings, royal, fondant, butter, frosted or glacé icing. Specify the advance notice needed for the more complicated cakes. Give details of the sizes and weights of the cakes available as well as the number of portions each will serve. Draw up a separate price-list so that this can be updated easily and remember to cost in extras such as boxes, bases and ribbons.

Flower arranging

This work can also be contracted out to either freelance florists or local businesses. Research into the standard of work of potential suppliers before putting them on your books: do they use only the freshest of flowers, can they work with a variety of materials, do they offer a wide range of displays, do they have a comprehensive stock of vases, bowls, baskets and plinths, how far will they travel, will they work weekends, how competitive are their prices? If possible, talk to people who have used them in the past.

Once you have satisfied yourself on these points, ask for a catalogue and price-list illustrating the range of their work – buttonholes, headdresses, bouquets, sprays, garlands, window boxes and tubs, arrangements with dried or artificial flowers. Many orders will be quite straightforward and simply a matter of the client choosing from the catalogue. Where elaborate and original arrangements are called for, however, the florist should meet the client to discuss things in detail such as themes, colour schemes and the number and sizes of arrangements.

Novelties and printing

Many functions call for novelty gifts or custom printing and you should be able to supply these items when called for. The sort of items that are likely to be in demand are small presents for place settings or lucky dips, crackers, snowballs, masks, streamers, novelty hats, balloons, sparklers and indoor fireworks, printed book matches, printed or novelty napkins, menus, programmes and invitations. Many of the novelties will be cheap, particularly if you are buying in quantity, but custom printing will not, especially if several colours or embossing and cut-outs are involved. Novelties can usually be ordered from suppliers' catalogues, but with printed items, orders will have to be discussed in detail with both the client and the printer. Proofs for all printing work should be cleared by the client in case any last minute alterations need to be made. Most people are not used to dealing with print work, so it is wise to warn the client not to make last minute changes unless absolutely necessary as these are always costly.

Toast-masters

A more accurate name for a toast-master is perhaps a master of ceremonies, for this is the role he plays at formal weddings and dinners. As well as announcing guests as they arrive and calling the toasts, the toast-master tells the speakers exactly what they have to do and when. He then takes charge of the events, seeing that each stage of the programme starts and finishes on time and that everything runs according to plan. Most speakers are nervous and a toast-master plays a very important role calming and reassuring everyone. Good toast-masters take a great pride in their work and will have gone through a rigorous training before embarking on their careers. They are experts in the etiquette and ritual of every type of function and will advise and even plan proceedings from beginning to end if the client is unsure of what is required. They should be booked well in advance to avoid disappointment.

Photography

It is impossible to over-emphasise the importance of using only the most reliable and professional photographers. Every function has a very special significance to participants or guests and a photographic record enables people to recall these once in a lifetime moments long after the event is over. Most photographers are reliable and skilled in their work, but there are those who may not turn up on the day or who are such novices that their pictures are as good as useless. Satisfy yourself thoroughly that the photographers on your books are totally reliable and book them early. If video is called for, remember that this involves quite different skills and techniques from still photography. Again, use only firms who specialise in this field and avoid the photographer who dabbles in the work. Make sure that the photographer meets the client before the event for a thorough briefing. If possible they should meet where the function is to be held, as lighting conditions and space may mean that the photographer has to bring along special equipment or that the spot the client has selected for photographs may be unsuitable and a new position needs to be chosen.

Function rooms

Some of the large catering companies have their own reception rooms which they rent out. Smaller firms who are unable to offer this facility may nevertheless be asked to recommend premises from time to time. It is worth compiling a comprehensive list of function rooms throughout your catchment area, calling to see the facilities and meet the people who manage them. If printed details are not available, make a note of the facilities, room dimensions and rentals and update these every year. Making yourself known to the management can be very valuable, as it not only means they will be more inclined to go out of their way to help you, but they may also pass your name on when there are enquiries for caterers.

There are numerous other ancillary services that a caterer may be called upon to provide – singers, bands, discos, comedians, dancers, magicians, Father Christmases, films and videos, coaches, singing telegrams, and so on. In all cases, reliability is the name of the game. Never employ anyone who has not been personally recommended or whose work you are not familiar with. And always make sure they are clearly briefed by either yourself or the client as to exactly what is required of them. In this way you will never disappoint a client and your reputation as a helpful and professional outfit will spread.

Chapter 18
Drink

Large catering concerns and companies at the top end of the market will employ their own specialists whose expertise in the buying, storage, pricing and serving of drink has been built up over a number of years. Employing experts of this calibre will obviously be out of the question for the smaller caterer. His turnover is unlikely to be large enough, and the demand for top quality wines and spirits too small to warrant employing such a specialist.

Nevertheless, the small caterer must be prepared to offer his clients a comprehensive service when supplying drink. It is a very lucrative service to offer because the mark-up on drink in this country is high and, if a caterer does not supply drink, he will lose business to those who do. Drink is an expensive item, one that is subject to deterioration, pilferage, fluctuations in price and quality, and no caterer can afford anything other than the most professional attitude to the buying, storage, selling and service of stocks.

Buying

The four main sources for buying drink are retail outlets, wholesalers, auctions and shippers. Each source has its advantages and disadvantages but in the final analysis it is probably the firm's turnover which decides where it buys.

Retail Outlets
Small caterers with a low turnover in drink will probably find they cannot afford to buy in bulk from wholesalers. They might be able to if they form a consortium and share orders with four or five other firms, but this is unlikely to work as their demands for product, quality and delivery would differ and rarely coincide. Buying wholesale quantities for themselves is likely to be out of the question, as storage would be

long term and therefore expensive, wastage might occur if stocks started to deteriorate and, of course, large sums would be tied up in stock when they could be more usefully employed elsewhere.

The most appropriate place to buy drink in these circumstances is the retail shop. Supermarkets and off-licence chains will offer the most competitive prices and the widest selection of well-known wines. Look out for bargain buys and 'own label' as these are usually excellent value. If you are a regular customer and get to know the people behind the counter, they may even give you advance warning of forth-coming bargain offers which can help you plan some of your buying. Ask their opinion of little known or new lines, as they are often genuinely interested in the business and take a pride in advising customers.

The small specialist off-licence will not normally be able to compete pricewise with the giants but this does not mean to say they should be ruled out altogether. Sometimes they will be the best source of fine wines and spirits. They may specialise in particular lines, and thus for certain functions they would be a more suitable place to shop than the chain stores. Also, the small retailer occasionally offers larger discounts per case than the bigger stores, so the price difference between the two may be less than it seems at first glance. Remember to check discounts and compare the difference from store to store and make enquiries about account facilities. If you are going to be a regular customer the shop should be prepared to offer you an account and this can save you time and money, cutting down the number of cheques you write, and reducing book-keeping entries and paperwork. An account will also give you a month's credit, sometimes more.

Wholesalers
For companies with a faster turnover and a little more storage space, the wholesaler makes the most suitable source of supply. While prices will not be as cheap as buying direct from the producer, they should be lower than chain store and supermarket prices. Cash and carries can be included here as their prices are usually equivalent to the wholesaler's, but do check this point as sometimes they can be more expensive. Wholesalers don't usually set any minimum purchase rules.

Their pricing is generally worked out on a banded basis, for example:

£xxxx)   for 1–5 cases
£xxx) per case for 6–11 cases
£xx   )   for 12–17 cases.

Alternatively, they may offer a discount based on the number of cases bought or it may be up to the client to negotiate what he considers a fair price for the order. Wholesalers will either deliver, which obviously saves the caterer time and money, or goods can be collected after phoning or sending through an order. There are wholesale outlets across the country though, obviously, caterers who are based in rural areas will have less selection and further to travel than town- or city-based caterers.

Whether the goods are being collected from the cash and carry or wholesaler, or delivered to your premises, remember to check the quantities both at loading and unloading stage. Where drink is concerned it is impossible to be too careful about security because there is so much pilfering.

Auctions

People who buy from auction sales tend to buy in large quantities, or small supplies of extremely fine wines and spirits. Auctions are thus suitable either for the large catering companies or for those dealing with the top end of the market where there is a call for the finest goods.

One of the main advantages of buying at auction is that wines and spirits can be bought at very competitive prices. A disadvantage is that you can never guarantee being successful. The larger auction houses have wine consultants who will be happy to give advice about the various lots being auctioned. Some have tastings the day before. Inexperienced buyers should be wary of buying without tasting or without advice, as all goods are bought 'as seen' and no auction house will take back a purchase.

There is an art to bidding which, for the inexperienced, can take a little getting used to. The bidding can move at a tremendous speed and the novice may either miss a bid because he is not fast enough or bid above his limit before realising what he has done. It is a good idea to visit a couple of sales just to get used to this before going in to bid yourself. If you are unable to attend on the day you can leave your bids with the auction room staff who will act on your behalf. Different auction houses have different fees which they add to the sale price. Remember to calculate these into your bidding as they can increase the price quite considerably.

While the main auctions are held in London, sales are held all round the country throughout the year.

Shippers

Tremendous savings can be made by buying direct from the producer. The drawback here is that the buyer has to arrange the shipping himself and becomes involved in time-consuming and fiddly paperwork. The solution is to buy direct from the producer, but through a shipping agent. The agent looks after all the shipping arrangements and documents and the buyer simply pays him for this service. The agent's fees are generally reasonable because he has no money at stake in the deal.

Storage

There are two aspects to consider about storage, the economic and the technical.

On the economic side, a caterer must weigh up carefully the price he is paying for the stock against the storage costs from the date of purchase to the day he sells it. Storage is an expensive commodity and unless a firm has its own storage space to spare, it is not usually a viable proposition. Rents for the storage space are likely to cancel out any savings made on bulk purchases or on new wines that have to be laid down. And even if the firm has its own storage space, it must consider carefully whether it is sensible to tie up a large amount of its money in stock when it could possibly be used more effectively elsewhere.

On the technical side, the most important factor is to keep the temperature steady - 55°F is a good level to aim at. The worst possible conditions are when the temperature fluctuates, as this puts stresses and strains on the drink. A temperature that is too hot speeds up the maturity rate so will bring stocks on too quickly. Too low a temperature will slow maturity down. Both could cause problems if stock had been laid down for a specific date. Wine must be stored on its side to prevent the corks from drying out. Spirits must be stored upright; if they are laid on their sides the metal caps will eventually rot.

Pricing

The formula used by most caterers for pricing wines is fairly straightforward. For quick turnover stock it is normal to double the purchase price and add VAT. If the wines have been laid down for some time, add 10 per cent and compound this each year, then in the year of sale, double the total and add VAT.

For spirits the formula is less rigid and it is best to see what sort of prices the competition is charging.

A mixed bar can be a headache for the small caterer. Larger companies who provide this service on a regular basis can simply count the number of tots consumed and charge the client accordingly. Small companies who are not often called on to provide a mixed bar will probably have to buy in stocks especially for the occasion. They obviously wouldn't want to be left with a whole lot of half full bottles, so the accepted practice here is to charge the client for all the bottles which have been opened and to take back all the unopened bottles. The opened bottles are handed to the client at the end of the function.

Corkage is charged when the client provides the drink himself. In theory, the corkage fee is the normal profit margin a caterer would expect to make on each bottle. In practice this is a bit steep, especially as the caterer has made no financial investment in the drink himself. The best thing to do is to have a look at what the competition charges and fit in with this.

Serving

Serving and presentation is as important with drink as it is with food. Wines must be served at the right temperatures and in the correct glasses. Cocktails must be mixed in the correct proportions and with the right ingredients. Drinks must be presented with the appropriate trimmings: olives, cherries, lemon slices, orange slices and so on. Whether your bar staff are employed on a full- or part-time basis, you must be certain that they are thoroughly trained in serving drink and that they also have a conscientious and professional attitude towards their work.

Be sure, too, that they are honest and, even then, still make strict security arrangements for the drink at all stages: when it is in transit, when it arrives, when it is packed up and when it is returned to the storeroom. Check quantities at each of these stages and when packing up at the end of a function, remember to check the empties as well as the full bottles. One of the oldest tricks in the book is to turn full bottles upside down and place these in a case of empties. Beware, too, of bar staff who are fond of their drink. Sadly, it is one of the hazards of the business that staff often become heavy drinkers or even alcoholics and help themselves to whatever is around. Not only are they then drinking your profits or the client's property, but they may also work less efficiently and even worse, they may get drunk and cause a scene.

Whether you are providing wines, a mixed bar, cocktails or punches, make sure you get an accurate briefing from your client. Some clients will know exactly what they want while others will need guidance. If there is any doubt over the choice of brands, get the client to try samples before finally deciding – it is well worth the trouble and cost to get it right on the day. The sort of points to cover in your briefing are:

Wines

Quality	vin de table
	medium quality
	fine wines
Type	red, white or rosé
	sparkling
	dry, medium or sweet
	spritzers
Countries	European
	New World

Should a selection be served?
If so, in what proportion?
Should different wines be served with each course of the
    menu?

Mixed bar
selection list:
Sherry - dry, sweet
Port
Gin
Whisky
Brandy
Vodka
Rum
Vermouth - red, white, rosé
Mixers

Chapter 19
Advertising and Public Relations

Activities aimed at bringing in the business are of paramount importance when you are just starting up. No one has heard of you or used your services and your first job will be to rectify that situation. Even if you have been operating successfully for some time you cannot just sit back and expect the business to continue coming in. Customers can be very fickle. They may decide to try another firm and then stick with them for a while, they may move away or they may just become bored with what you have on offer.

Advertising and public relations activities are aimed at achieving a continuous flow of new business. Depending upon how successful your efforts are you may find that you will need to regulate expenditure in these areas to fit in with seasonal differences in the flow of work or to coincide with the introduction of new services or products. The programme will also need to be dovetailed with plans for direct activity such as mailings and personal visits.

In fact, the main difference between advertising and public relations is that advertising deals with paid-for space in the media ie, newspapers, radio and magazines, whereas public relations is concerned with persuading opinion formers such as newspaper editors, radio presenters, business entertainment organisers and indeed each and every customer to write or talk about your business and its services. Do not underestimate the last item for word-of-mouth recommendation is probably the most important way in which your business becomes known.

Choice of media

To be effective, advertising and public relations activities must be directed towards the right audience. There is no point, for example, in advertising in *Tatler* magazine if you can only cover a very small area and are specialising in children's parties. Equally, a write-up in the local county magazine will not be very helpful if you want to concentrate on business luncheons.

Take another look at the business proposals which came out of Chapter 1. What kind of services are you planning to offer and who is likely to be interested? The answers to these questions will marry up your plans to the media which these people are likely to take. Children's party givers might well read county magazines or the *Lady*. The businessman, on the other hand, may be more usefully approached through the mail.

Quite often very useful sections of the media get overlooked. Office notice boards, for example, can be a very effective means of local communication, as can theatre programmes or even racing cards. Some media are really only open to paid advertising. Others will be receptive to PR approaches for a simple listing in the shopping column or perhaps for a full-scale interview.

Here's a run down of the kind of media which you should consider, together with an indication of the sort of material they might use. You will have to check for yourself what kind of people receive them and how many.

Television
For most small to medium firms, television advertising is out of the question for it is too expensive. However, the advertisement rates on Channel 4 are nothing like so high as others and may be of interest to larger companies with specialist activities. Each independent television area has separate rates so you need not advertise nationally. TV programme coverage is unlikely but, if you have an outstanding story to tell of rags to riches or the provision of particularly unusual services, they could be interested. Try the editors or producers of the local feature programmes. However, this would be very much a one-off item.

Radio
This is likely to be a much more useful medium for the smaller business. Rates are not very high and most independent local radio stations have facilities for recording simple advertisements. Make sure that the station's catchment area matches your own or you may find yourself travelling rather further than you had planned.

This is also an excellent target area for public relations activities. Most stations – both independent and BBC – are on the look-out for local people to interview. All you need to do is think of an interesting angle. If the interview goes well you might be able to organise a regular spot giving hints and tips on easy entertaining, mixing cocktails or some aspect of specialist cookery. Make sure that you are identified by your business as well as by your own name.

Newspapers
Advertising in national newspapers can be a costly business but it may be a good way to get to your particular target audience. Remember, there are always the personal columns as well as the display ads and these can be much more reasonable. However, if you live outside London and are limited in the distances you are prepared to travel, regional or local newspapers may be a better bet.

Check the circulation areas of the newspapers which serve your catchment area and find out what sort of people read them. Daily papers may have larger circulations than weeklies but they will also be more expensive and have a shorter life. Weekly papers are often kept for the entertainment pages. The best place in the newspaper for your business may be the entertainment pages but it could equally be the business section.

Editorial coverage will be extremely useful in the areas where you are operating. Always send out a press release (see page 152) when you open up and when you offer a new product or service. Some of the jobs you do may also be newsworthy. For example, you may be doing the catering for a big opening or for a VIP visit. If so, ask the people who are organising the publicity for this event to include your business in the write up, and if any reporters attend, give them the details yourself. If you have an uncommon speciality, the local paper may be interested in your story. Approach the person who runs the home page or the shopping column, or failing this, ring the features editor.

Magazines
Both the advertising and the public relations possibilities on magazines are rather similar to those of newspapers. However, magazines have a longer life and they also tend to be aimed more specifically at similar interest groups. In addition

to the many national magazines, there are county and local town magazines. These often concentrate on leisure activities and are usually bought by those with a little extra money to spend. House magazines produced for the employees of local firms may also be a useful place to advertise.

Yellow Pages and local directories
The advantage of directory advertising is that it is very local. If you have a business telephone connection you will automatically get a free line entry in your local Yellow Pages. If you are operating from home it is worth making sure that you change over from a private connection. The rental is a little more expensive but the Yellow Pages entry is worth it. An advertisement will, of course, make your entry stand out and the Yellow Pages staff are quite good at helping to draw up a small display ad.

There are also some other reputable local directories but, if you are approached by a local directory salesman, make sure his directory really does exist for there are spurious ones around. There are also some organisations which specialise in solving people's problems and it could be a good investment to get on to their lists too.

Programmes
If you have a local theatre, a special sports stadium or race course, it could be worth advertising in these. Theatre-goers may well be heavy entertainers and race-goers could be in the market for special packed lunches. County shows, horse trials and motor racing offer other opportunities. The decision to advertise should depend upon your assessment of how many people at the particular event in question are likely to be interested in your kind of services.

Notice boards
These can be a very useful form of communication and often a very reasonable one. If your business is small scale and local in its operations, use shop window boards where it will not be difficult to make your card stand out. Club and office notice boards can be equally effective at bringing in business. Approach the club secretary for the former and the personnel manager for the latter. Keep your card small and neat - it should not take up too much space on the board.

Direct mail

With direct mail your message goes straight to the home of the private client or to the desk of the business client. Compiling a list of suitable addresses may take a little time but the local Yellow Pages, engagement announcements in the local papers and street directories can all help. You could also have your message hand delivered to houses in areas which your market research indicates to be filled with potential customers. Write a covering letter setting out details of your services and enclose any suitable promotional literature.

The tools

To be really effective in the advertising and public relations field it is not enough to choose the right media outlets, you must also get your message across. This means that you will need to give some careful thought to the content and design of your advertisements and direct mail shots and to the content of any press release.

The secret is not to try to be too clever. Stick to the point in all your promotional material and simply set out the details of the services you have to offer with reference to the way in which they can make life easier and pleasanter for the recipients.

Advertisements

The telephone sales people on newspaper and Yellow Pages advertising desks are often trained to help in putting together simple advertisements. Nevertheless, it is sensible to have some idea of what you want to do. Take a look at competitive advertisements and see which ones strike you as being the most effective. Good clear lettering stands out better than script or italics and this, coupled with a simple drawing, may be the answer. Make sure that the name of your business is easy to read and that the entry includes your telephone number and address. A simple line such as, 'Let us arrange your Party for all Special Occasions' could sum up your business very well. However, if you do have any special services to offer such as vegetarian cuisine, special theme cakes or barbecues, put that in too, but take care not to overcrowd the advertisement. Take a larger space if you have a lot to say.

Direct mail letters

Once again the key is brevity. The recipient does not want to read long eulogies about your own company. All he might want to know are what services you have to offer. You can always extol their virtues in greater detail if you get an enquiry.

The contents of such letters must, of course, relate to the recipient so that a letter to potential business customers will be rather different from a letter to a club secretary.

Press releases

These are really quite easy to write. Stick to the facts and resist the temptation to pad things out. Get as much information as you can into the first few sentences or paragraph and develop the more interesting points in subsequent paragraphs. This makes it much easier for an editor to cut to a particular length or to lift out sections to fit in with a shopping column or services round-up.

Send out a release when you open up and whenever you have new services to offer or when you have fulfilled a particularly interesting or unusual job. Send a photograph, if possible, with the latter. Remember to add the contact details at the base of every release sent out and fix the release firmly to the picture.

Here's an example of a starting-up release. Sample menus should be included with such a release.

NEW CATERING SERVICE FOR SUSSEX
Barbecues, Picnics & Outdoor Food a Speciality

A new catering service specialising in outdoor food of all kinds is now on offer from Teddy Bears' Picnics of 6 Alton Road, Farmbridge, Sussex; Phone 0000 11111. Orders can be delivered any time, anywhere within the county boundaries.

The barbecue service includes preparation on or off site, delivery of the food and clearing away. Service, equipment hire, special theme/support and drinks can also be arranged.

Picnics and special hampers for Glyndebourne and other artistic and sporting events in Sussex are another speciality of the new service. Prices range from £5 per head for simple picnic baskets to £15 plus for special events hampers.

Teddy Bears' Picnics also offer indoor party catering, specialising in theme birthdays and anniversaries and children's parties.

For further information contact:
A Anon, Proprietor
Teddy Bears' Picnics
6 Alton Road,
Farmbridge, Sussex
Tel: 0000 11111

Chapter 20
Dealing with Clients

Your day-to-day dealings with customers and clients can be equally important in keeping the business flowing in. Indeed, in some ways this is just as much a part of your public relations programme as sending out press releases or talking on the radio. A happy and satisfied client is one who is likely to come back to you, but a disgruntled one will not and he may even put other potential customers off.

Dealing with people is not always easy and it can take much patience and tact to handle someone who seems determined to be difficult. Keep the old adage that the customer is always right to the forefront of your mind – even when he is not! No one will thank you for showing him up or proving him wrong, so tread carefully even when you know you are in the right.

A firm and friendly approach is equally important and you must appear to know what you are doing. After all, a customer needs to believe that you are an expert in your field and will not let him down on what to him might be an extremely important occasion. In short, you should try to project an image of competence and reliability, and the best way of doing this is to *be* competent and reliable. Don't make promises you cannot fulfil and avoid the temptation to take on business when really you are fully stretched.

Starting the dialogue

The first step is to have an efficient and helpful answering system for enquiries. If you are a small one-man band this may mean a really friendly and welcoming message on the answerphone machine. This is very important as there are lots of people who do not like speaking to a machine and feel intimidated by an abrupt instruction to talk after the bleep.

Some small firms opt for a personal answering service. This may be on a separate number or linked to one or more of your own numbers. Larger firms may have receptionist/telephonists who take the calls initially in the absence of more senior management. In both these circumstances it is a good idea to ring up yourself to see what sort of greeting and response are given, and ask some friends to ring in with fictional enquiries and to make notes on how they are dealt with.

Once contact is made between the potential customers and the person who will be organising the event, the emphasis must be on eliciting as much information as possible. It is important both for the customer and for your planning to get all the facts at the start.

Some customers are not really sure exactly what they want and the initial conversation should be aimed at helping the client to make up his mind without pushing him into something that he might regret afterwards. Other customers seem to know exactly what they want and though this may be a help, it can also have its drawbacks. The potential client's ideas may not be practicable, they may be cutting the corners on staff and they could be contrary to the knowledge which you have of what makes a successful event. If this is the case you will have to use a good deal of tact to change the customer's ideas without antagonising him.

The dialogue with the customer should continue during the planning stages for an event through to the day itself. You ignore your customer at your peril. Communication at crucial stages ensures that you are all thinking along the same lines and helps to eliminate the possibility of misunderstandings. It also reduces anxiety on the part of the client which makes life easier for all concerned.

It should go without saying that you must listen to your client but some caterers are surprised that, at the end of the day, their client is not happy with the arrangements even though there was consultation at every stage. Of course, there are occasions when it seems that customers are being unreasonable, but check your own notes before blaming them.

Problem customers

There should not really be any of these if you are doing your job well but some people can be difficult, sometimes, it seems, just for the sake of it! Every client is a law unto him or herself but here are a few typical problem customers with some tips on how to deal with them.

The vacillating customer can drive you round the bend with continual phone calls changing some aspect of the plans for his event. First he will change the menu, then the time the food is to be served, then the drinks and then the menu again. He will order flowers and cancel them and then blame you if there are no floral decorations on the day. The only way to deal with this kind of customer is to be as firm as possible and to try to discourage too many changes. Confirm in writing the changes he does insist on so you cannot be blamed for them later.

Most customers tend to worry about their event, but they worry particularly about the time you are to arrive and about the quantity of food. Regarding the time of arrival, do keep to what was agreed in the first instance, or if you decide that you do not need to arrive so early, ring up and let your customer know. You will get off to a bad psychological start if the customer is in a panic over your non-arrival. With regard to the quantities, things can be the other way round so don't let your customer panic *you*!

You may have some customers who do not know how to treat staff. Of course, you will not know this the first time you cater for them but on a subsequent occasion you can alert your waiting staff and assistant cooks to the fact that the client may be rather off-hand or that he is particularly exacting or difficult. Forewarned is usually forearmed and you do not want your staff leaving you because of the actions of one client.

Dissatisfied customers, on the other hand, should be listened to with some care. Encourage customers to make their complaints direct to you, but do also make sure that your staff come to you with any complaints from customers, then check things out in detail. Once you have the facts you will be in a much better position to deal with the complaint. Direct apologies may be needed; if so do give them personally and in writing, possibly with some offer to make amends. Even if you do not think that your operation is at fault, it is worth listening carefully to complaints for they could arise from misunderstandings or simply from bad communication. You may find quite simple remedies to ensure that the same thing does not happen again.

Customers who are late in paying or who do not pay at all are the worst of the problem customers. You must first make sure that each customer has been informed of and understands the terms of payment. The next step is to send out reminders promptly. Do not let things slide; everyone tends to pay those who press for payment sooner than those who do not follow up very quickly.

If you still do not get paid you may have to send a solicitor's letter or even resort to the courts, which does not have to be too costly. There are various procedures which you can put into action without any specialist legal advice and these are worth finding out about. Talk to your local TEC and go to the local law centre. Some private firms even run courses on efficient debt collecting.

Follow through

Good after-sales service in terms of clearing away, collection of hired equipment, the despatch of wedding cake and the like all help to create a good impression with customers. It is also worth checking to see if there are any other events linked to the main one which might require catering.

A general follow-up letter after the job has been completed is a good idea and this might be followed up with six-monthly or yearly contact to see if any other major events are to be planned: every client is a potential customer for your mailing lists. Clients who have arranged Christmas parties with you should be contacted in good time for the following year's event, wedding breakfast clients could become dinner party or buffet party customers and children's party customers may be interested in summer picnics or barbecues.

A Christmas card featuring one of your latest menus is another way of keeping in touch and you will probably be able to think of some more creative ideas.

Chapter 21
Adding a Restaurant

Though running a catering company and running a restaurant are both forms of catering they do involve quite different considerations. Some successful catering businesses are run from restaurants but apart from possible overlaps in the kitchen the control and accounts systems are best kept separate.

Location and premises

The first major difference lies in location. You could quite easily run your catering business from a quiet cul-de-sac or a remote village in the country but these would obviously not be very good sites for a restaurant. Instead you must be close to your customers and this will mean the High Street or maybe proximity to new suburban shopping malls. Remember that the 'centre' of a town can change and your restaurant needs to be where the people are.

If you find premises you like, look at the other businesses around to assess the market potential. A McDonald's means plenty of people and a Marks and Spencers spells a certain social mix, probably with some spare cash for eating out. Don't worry if there are other catering establishments nearby. Provided you are not offering exactly the same type of mix a little competition can be very healthy. You may also gain customers from the existing trade.

You should also consider whether to buy a going concern or to start from scratch. If you buy a going concern you will not have to worry about planning permission; on the other hand, you may be buying somewhere with a bad reputation. It is important to do a bit of local research to find out how the place is regarded and also to try and find out why the current owner is selling. He may know something about the area that you do not. The residential status of the area may be changing or a large local employer may have closed, both very good reasons for not buying into the area.

If you are planning to start from scratch you will probably need to find new premises. You can operate your catering business from a nissen hut or premises under the railway arches but restaurant premises will need to have a much better image. The building will need to be in much better condition both inside and out. This is not to say that the premises for your catering company reach any lower hygiene standards but that a restaurant must not only be clean and hygienic but look the part too.

The restaurant and kitchen must be at the right level. Provided that there is a lift it will not matter to your catering company if your kitchen is on the fifth floor but this will not do for a restaurant. You may have to do some extensive re-structuring, so you will need a surveyor to check the building is strong enough to take the changes planned and that the gas and water systems are in the right place for the kitchen, and staff and customer toilets. Indeed, you should also hire a surveyor to check any existing premises you are thinking of buying.

If you are planning to set up the restaurant in your own premises or in an extension of any kind, consult the district surveyor at once. Increasing the number of people who are visiting a building may have implications for Health and Safety Regulations and for Fire Regulations.

The atmosphere and decor of a restaurant need careful planning too. The atmosphere of a place can quite literally mean the difference between success and failure. Too cold and clinical is just as bad as cluttered kitsch. Your restaurant needs to have a deliberate theme or feel to it which extends from the name, through the decor to the menu and food.

Food and drink

The food you serve in the restaurant may be similar to the menus you offer for outside catering but the considerations are very different. With outside catering you know exactly how many people will be eating which dishes. Even if you have a set, no-choice menu (and this is not very popular) in a restaurant you do not know exactly how many people you will be serving. With the more usual multiple-choice menu you do not know how many people will choose the Coq au Vin, nor how many the Guinea Fowl with Peppers.

However, there are some guidelines; the *Caterer and Hotelkeeper* magazine publishes an annual list of the most popular restaurant dishes and though you may not be planning to serve Prawn Cocktail, Grilled Steak and Black Forest Gateau (consistently at the top of the list) you will know that steak sells, as does any kind of grilled or fried chicken, deep fried mushrooms and fish in batter. Experience will also help you to decide how much of each of your dishes you will need to have in reserve. Because of this problem it makes sense to develop 'dishes of the day' which use up leftovers.

There may be some overlap of menus from the two operations and batch baking for catering can also produce pies and puddings for the restaurant. The restaurant might act as an advertisement for the catering company's food and vice versa though you do not want to carry this too far or you will lose flexibility. Some dishes will be appropriate for both operations, others will not.

You will also need to look carefully at your kitchen and hygiene control systems. Changes may be necessary with the addition of new equipment and a larger number of kitchen staff.

If you have been buying wine on an *ad hoc* basis for the catering company you will need to completely rethink your strategy, for the restaurant will need a wine list. If you have already started to buy wine in quantity things will be easier but you will still need to work out exactly what will sell well in the restaurant and what will be best for outside catering.

Unless you know a lot about wine it makes sense to work fairly closely with one or two local suppliers. They will be able to advise you and may even hold some of your stock for you if you do not have much storage space in the restaurant. You should be able to negotiate good prices and build up a personal relationship with your suppliers.

If you do not have any contacts in this field talk to as many merchants as you can and listen carefully to what they have to say. Beware the salesman who starts by telling you how reasonable the wine is, follows this up with a eulogy on the presentation and talks about the contents of the bottle last. The order should be the other way round!

It is not necessary to have a very extensive wine list. A

choice of 8-12 white wines and a similar number of reds should give you a good range. Do try to include some half bottles and choose a decent house wine to serve by the glass. Cutting corners here is shortsighted. The majority of customers choose the house wine and you will be judged on it.

Nor does the wine list need to be elaborate, but do take the trouble to include vintages and the name of the grower, negociant or supplier. This helps the customer to make a choice and shows that you know what you are doing. You will also need to have a well stocked bar even if this is not set up in the restaurant. You will find that you can make good profit centres out of pre- and post-dinner drinks.

A wine list and a bar of course mean that you have to have a licence. This is not necessary for a caterer unless you are operating a cash bar at a public event. The licence must be in your name, so even if you are taking over an existing restaurant you will need to re-apply for the licence. This is not usually withheld unless there have been a lot of complaints about the premises in the past. Check this out before you buy.

New premises mean a new licence and these can sometimes be difficult to get. Very often local licensed premises will object to another licence in the area. However, this usually applies to full licences under which you can serve any kind of alcoholic drink without serving food. You may prefer to go for a restaurant licence which allows you to serve all kinds of alcoholic drinks provided customers are having a meal, for this is often easier to get.

Control systems

Stock control becomes even more important if you are running a restaurant as well as a catering company. But you will need to balance the advantages of negotiating larger discounts for food and wine bought in larger quantities against capital tied up in that stock. Larger quantities of stock also offer greater opportunities for pilfering. With a catering company you might be the only person handling expensive stock. With a restaurant there will most certainly be others.

This also applies to the money. Waiters and waitresses will be taking the cash and even if they work through a cashier you should give some thought to choosing honest staff and working out ways of detecting dishonesty if you have been unlucky. Computerisation and electronic tills may be a first step in this direction.

This kind of equipment will also help you to separate the accounts of the various aspects of the business so that you can see how each is working and how they are responding to any changes that you make.

Running a restaurant will also change the pattern of staffing and the chances are that you will be employing many more full-time staff than before. With 16 or more Acts of Parliament covering employment this is the time to look at your employment accounts.

You may also need to review the way in which you recruit staff. Many of your new staff will be dealing with your customers. They are your ambassadors and sales force. In the restaurant people are coming to you and they must be able to feel welcome. They must also be looked after and cared for. Within this framework your staff can also improve your profits by selling the dish of the day, pre-dinner drinks, mineral water, wine and coffee. In these circumstances you might consider starting some form of incentive scheme for your staff.

Keeping the restaurant full

It is not enough to get the restaurant going – you have got to keep it full. Novelty value may bring the customers in to start with and a good menu and efficient, friendly service will keep them there – but only for a while. It is fatal to sit back thinking that things are going well. You must work at keeping people interested. If you do not they will start to drift away.

The menu is the obvious area for change. The same menu will start to bore after a month or two. On the other hand, if you change too much customers will start to complain that you have removed their favourite dishes. A combination of set seasonal menus, changing four times a year, and a reasonable list of dishes of the day could give you the right combination.

Flexibility is important. Food fashions change and so do the people living in your area. Make sure you are keeping your finger on the pulse of what's happening around you. It may be that you could fill in quiet times with private parties brought in by your catering company or you might organise special events offering regional cuisines prepared by a guest chef.

Special events and public relations are useful tools but

word of mouth recommendations are the best. People will recommend your restaurant if they have enjoyed themselves but they will also be very quick to condemn it if they have a single bad evening.

The answer is to set a standard and to keep to it. Are you all as genuinely welcoming as you were when the restaurant first opened? Do you take time to talk to customers and see how they are reacting to your menus? Is the food preparation and presentation as meticulous as it should be? Check yourself and your staff every day.

Appendix

Catering and cookery courses

- ● Offers short courses: weekends, full week, single day or one day for a number of consecutive weeks.
- * Offers a three-month or longer diploma course.
- ○ Offers two- to four-week courses.

- ● **Aga Workshop**
  Mary Berry, Watercroft,
  Church Road, Penn, Buckinghamshire HP10 8NX
- ● * ○ **The Ballymaloe Cookery School**
  Kinoith, Shanagarry, Co Cork, Eire
- ● ○ **The Bath School of Cookery**
  Bassett House, Claverton, Bath,
  Avon BA2 7BL
- ● **Betsy's Kitchen**
  3 St James's Gardens, London W11 4RB
- ● ○ **Bonne Bouche School of Cookery**
  Lower Beers, Brithem Bottom,
  Cullompton, Devon EX15 1NB
- ● ○ **Cameron Cooking**
  32 Rusholme Road, Putney,
  London SW15 3LG
- ● **Catherine Blakeley's Cookery Courses and AGA Owners Club**
  Arlington House, Station Road,
  Newport, Shropshire TF10 7EN
- ○ **Cookery at The Grange**
  The Grange, Whatley Vineyard,
  Whatley, Frome, Somerset BA11 3LA
- ● * **Cordon Bleu Cookery School (London) Ltd**
  114 Marylebone Lane, London W1M 6HH
- ● **La Cuisine Imaginaire**
  Roselyne Masselin's Vegetarian
  Cookery School, 4 Blenheim Crescent,
  Notting Hill Gate, London W11 1NN
- ● **The Earnley Concourse**
  Earnley, Chichester, West Sussex PO20 7JL

* **The Eastbourne College of Food and Fashion**
1 Silverdale Road, Eastbourne,
East Sussex BN20 7AA

● * ○ **Edinburgh Cookery School**
The Coach House, Newliston,
Kirkliston, Edinburgh EH29 9EB

* **Egglestone Hall**
Talbot Gray Ltd, Barnard Castle,
Co Durham DL12 0AG

● **Farthinghoe Fine Wine and Food**
The Old Rectory, Farthinghoe,
Brackley, Northamptonshire NN13 5NZ

● **Frances Kitchin**
Stoneymead, Curry Rivel, Langport,
Somerset TA10 0HW

● **The Horn of Plenty Restaurant**
Culworthy, Tavistock, Devon

● **Jill Probert's Cookery
Demonstration Courses**
Bretton Hall, Chester, Cheshire CH4 0DF

● * **Leith's School of Food and Wine**
21 St Alban's Grove, London W8 5BP

● **Manor House**
Chenies, Rickmansworth,
Hertfordshire WD3 6ER

● **Master Classes in Food and Wine**
27 Gibwood Road, Northenden,
Manchester M22 4BR

* ○ **Elisabeth Russell**
Flat 5, 18 The Grange,
Wimbledon, London SW19 4PS

● **Squires Kitchen Sugarcraft**
Squires House, 3 Waverley Lane,
Farnham, Surrey GU9 8BB

* **Tante Marie School of Cookery**
Woodham House, Carlton Road,
Woking, Surrey GU21 4HF

● **The Vegetarian Society
Cookery School**
Dunham Road, Altrincham, Cheshire WA14 4QG

Useful addresses

National telephone dialling codes are given, though local codes may differ.

Local councils, tourist boards and chambers of commerce can be good sources of help and information. Many organisations listed below will have local offices.

Action for Cities Coordination Unit
Room P2/101, 2 Marsham Street, London SW1P 3EB; 071-276 3053
Advisory, Conciliation and Arbitration Service (ACAS)
Head Office, 27 Wilton Street, London SW1X 7AX; 071-210 3000
British Hospitality Association
40 Duke Street, London W1M 6HR; 071-499 6641
Development Board for Rural Wales
Ladywell House, Newtown, Powys SY16 1JB; 0686 626965
Department of Employment
Information Branch, Caxton House, Tothill Street, London SW1H 9NF; 071-273 3000
Department of Health
Alexander Fleming House, Elephant and Castle, London SE1 6BY; 071-972 2000
Department of Social Security
Richmond House, 79 Whitehall, London SW1A 2NS; 071-210 3000
Employment Department Training and Enterprise Councils (TECs)
England and Wales: Freefone 0800 444246 for all regional offices or ask at your Job Centre
Enterprise Initiative
Freefone 0800 500200 for regional addresses throughout England
Wales: Welsh Office Industry Department, New Crown Buildings, Cathays Park, Cardiff CF1 3NQ; 0222 823976
Scotland: Industry Department for Scotland, Alhambra House, Waterloo Street, Glasgow G2 6AT; 041-248 4774
Ireland: Department of Economic Development, Netherleigh, Massey Avenue, Belfast BT4 2JP; 0232 763244
The Federation of Small Businesses
5th Floor, 114 Union Street, Glasgow G1 3QQ; 041-221 0775
The Food Safety Directorate
Ministry of Agriculture, Fisheries and Food, Ergon House, c/o Nobel House, 17 Smith Square, London SW1P 3JR; 071-238 6550
Health and Safety Executive
Baynards House, 1 – 3 Chepstow Place, Westbourne Grove, London W2 4TS; 071-229 3456
HM Customs and Excise
VAT Administration Directorate, New King's Beam House, 22 Upper Ground, London SE1 9PJ; 071-620 1313
HMSO Publications Centre (enquiries)
PO Box 276, London SW8 5DT; 071-873 0011
Hotel and Catering Industry Training Company
Head Office, International House, High Street, Ealing, London W5 5DB; 081-579 2400

Hotel, Catering and Institutional Management Association
191 Trinity Road, London SW17 7HN; 081-672 4251
Local Enterprise Development Unit
LEDU House, Upper Galwally, Belfast BT8 4TB; 0232 491031
London Small Firms Centre
11 Belgrave Road, London SW1V 1RB; Freefone 0800 222999
Registrar of Companies
Companies House, 55 City Road, London EC1Y 1BB; 071-253 9393
Companies House, Crown Way, Maindy, Cardiff CF4 3UZ; 0222 388588
Companies House, 102 George Street, Edinburgh EH2 3DJ; 031-225 5774
IDB House, 64 Chicester Street, Belfast BT1 4JX; 0232 234488
Rural Development Commission
141 Castle Street, Salisbury, Wiltshire SP1 3TP; 0722 336255
Scottish Business Shop
120 Bothwell Street, Glasgow G2 7JP; 041-248 6014
Scottish Enterprise - Local Enterprise Companies (LECs)
120 Bothwell Street, Glasgow G2 7JP; 041-248 2700
Welsh Development Agency
Pearl House, Greyfriars Road, Cardiff CF1 3XX; 0222 222666

Further Reading

Financial Management for the Small Business: The Daily Telegraph Guide, 2nd edn, Colin Barrow (Kogan Page)
Law for the Small Business: The Daily Telegraph Guide, 7th edn, Patricia Clayton (Kogan Page)
The Modern Patissier, William Barber (Northwood Publications)
New Larousse Gastronomique (Hamlyn)
Best Wine Buys in the High Street, Judy Ridgway (Foulsham) published annually
Working for Yourself: The Daily Telegraph Guide to Self-Employment, 13th edn, Godfrey Golzen (Kogan Page)

Journals

Caterer and Hotelkeeper, Quadrant House, The Quadrant, Sutton, Surrey SM2 5AS
Catering, 161-5 Greenwich House, Greenwich High Street, London SE10 8JA
Decanter, St John's Chambers, 2-10 St John's Road, London SW11 1PN
Gourmet Magazine (US publication; can be ordered through newsagents)

Taste Magazine, Douglas House, 3 Richmond Buildings, (off Dean Street), London W1V 5EA

Which? Wine Monthly, Consumers' Association, 14 Buckingham Street, London WC2N 6DS

Wine Magazine, 60 Waldegrave Road, Teddington, Middlesex TW11 8LG

Index of Advertisers

Index